More praise for *Elephants on the Edge:*

"This book . . . is fascinating . . . [and] sheds light on disturbing phenomena relevant to the future not only of elephants but also of humans subjected to similar disruption. Read it."

—Robert M. May, Professor Lord May of Oxford OM AC KT FRS

"Groundbreaking and [a] remarkable feat of scholarship. . . . This fascinating book [shows] that we cannot understand the . . . relationship between man and other co-inhabitants of the natural world without insights into the deeper psychological and ethical substrata of our own minds."

—Allan Schore, Department of Psychiatry and Biobehavioral Sciences, David Geffen School of Medicine, University of California at Los Angeles

"Revolutionary and very exciting, [*Elephants on the Edge*] is important in terms of both elephant biology and elephant welfare."

—Cynthia Moss, Amboseli Trust for Elephants

"*Elephants on the Edge* is a wide-ranging, passionate, well-researched, and urgent call to action. These magnificent, intelligent, and emotional giants are quintessential poster animals for the wounded world in which we live. Read this book, share it widely, and please do something to increase our compassion footprint before it's too late. Healing demands collective cross-cultural action now."

—Marc Bekoff, University of Colorado, coauthor with Jessica Pierce of *Wild Justice: The Moral Lives of Animals*

"A poignant presentation of the eradication of elephant societies. . . . The arguments transcend the subject matter of elephants and herald a new cultural stance on human-animal relationships."

—Lori Marino, Emory University

"In *Elephants on the Edge,* G. A. Bradshaw helps us face our ethically flawed relationship with animals and nature and what is at stake for all of us."

—John P. Gluck, University of New Mexico; Kennedy Institute of Ethics, Georgetown University

YALE UNIVERSITY PRESS NEW HAVEN AND LONDON

Elephants on the Edge

What Animals Teach Us About Humanity

G. A. Bradshaw

Frontispiece and chapter-opening photographs courtesy of Cyril Christo and Marie Wilkinson.

Published with assistance from the foundation established in memory of Philip Hamilton McMillan of the Class of 1894, Yale College.

Designed by Nancy Ovedovitz and set in type by Technologies 'N Typography. Printed in the United States of America by Sheridan Books.

Library of Congress Cataloging-in-Publication Data

Bradshaw, G. A. (Gay A.), 1959–
Elephants on the edge : what animals teach us about humanity / G. A. Bradshaw.
p. cm.
Includes bibliographical references and index.
ISBN 978-0-300-12731-7 (cloth : alk. paper) 1. Elephants—Behavior. 2. Elephants—Psychology. 3. Elephants—Effects of human beings on. 4. Social behavior in animals. 5. Captive wild animals. 6. Psychology, Comparative. I. Title.
QL737.P98B73 2009
599.67'15—dc22 2009014004

A catalogue record for this book is available from the British Library.

This paper meets the requirements of ANSI/NISO Z39.48–1992 (Permanence of Paper).

10 9 8 7 6 5 4 3 2 1

To Fritzi

One does not meet oneself until
one catches the reflection from an eye
other than human.
Loren Eiseley

Contents

Illustrations follow page 114.

Foreword

Elephant breakdown, the subject herein, disturbs me. It says my own was inevitable. Recall Nietzsche's crackup, triggered by the sight of a tradesman flogging a horse, and you begin to understand what I'm talking about.

We are all susceptible. Descartes in dressing gown before his hearth, demolishing, as if brick by brick, his rational mind—one of the more famous crackups of history. The cloak of composure we wear carries its own unraveling—the bit of thread lying exposed. Sometimes, as with Nietzsche, it happens in a thunderclap of shattering dissonance.

Whether swift or slow, this is how we grow. "By being defeated, decisively, / by constantly greater beings."[1]

Thoreau's crackup occurred on a camping trip in the Maine wilderness. Alone in the swirling mists at Katahdin's summit, he felt a great energy moving near him. "What is this Titan that has possession of me?" *"Contact! Contact!"* he shouts into the pages of his journal. "*Who* are we? *Where* are we?" The sober author of *Walden* was in a place not controlled by man—untouchable, impenetrable, and impalpable. The mountain awakened with "a force not bound to be kind to man," Thoreau noted chillingly.[2]

So did the untouchable, impenetrable, impalpable Uncertainty that wrestled with Jesus of Nazareth for forty days and nights ("that kneaded him as if to change his shape").[3] Forty days in the presence of something language cannot reach—yet reborn, in the end, into the Jesus of History.

Carl Jung, in search of the primal mind on a trip through equatorial Africa, found himself perilously close to the same experience. "The trip revealed itself as less an investigation of primitive psychology . . . than a probing into the rather embarrassing question: What is going to happen to Jung the psychologist in the wilds of Africa?"

"The primitive," Jung realized, "was a danger to me."[4]

So it is to all of us, Dr. Jung. A scientific investigation into primitive psychology backfired into an explosive encounter with the First World of totemic consciousness—where consciousness is no longer exclusively human. In this eerie commonwealth of symmetrical, mirrored minds, the psychologist risked becoming something changed. Like Thoreau atop Katahdin, Jung found himself poised to *know* as he was *known*.[5]

The phrase is St. Paul's, and it is a bombshell. It marks the watershed separating *where the wild things are* from *where the domesticated things are.*

The First World bubbles just beneath the surface in each of us. C. G. Jung would agree. Jung would also agree that as children we are born into this chthonic consciousness (ontogeny recapitulates phylogeny), and, he would add, we ignore this deep, wild-visaged consciousness at the peril of our sanity. "Whenever we give up, leave behind, and forget too much, there is always the danger that the things we have neglected will return with added force."[6]

Elephant consciousness—human consciousness. Two sides of the same coin, says Bradshaw. Ever since Adam's Wall (beast on one side, mankind on the other), we have struggled mightily to suppress this cosmic law. (Daniel Quinn calls it The Great Forgetting.)[7] Until, like Nietzsche, we snap.

It is worth remembering that long before the Middle Eastern invention of sky gods, our ancestors followed labyrinthine chambers down into the entrails of the earth to stencil their hands on smoky walls—reaching across a membrane of stone to touch and be quickened by the spirit of animal kinsmen. Chauvet, Cosquer, Gargas, Pech-Merle, Cougnac. Subterranean halls whose walls rippled with bison, horses, oxen, lions. On the other side of the equator herds of eland, kudu, and springbok thunder by on open-air boulders in Namibia and Botswana. In Australia's Arnhem Land marsupial and man dream the same dream.

Ever since Lucy the Australopithecine, mankind has known that the coin of consciousness bears—must bear—two sides. Otherwise there is no viable or sane currency on earth.

Except for Adam's heirs.

My crackup happened on the day a Yup'ik Eskimo handed me a scrap of paper whereon was penciled, *"I am a Puffin!"* (I had just finished giving a class to a room full of Eskimos incarcerated in the regional prison, just below the Yukon River.)

He was not being charming. Nor poetical. Nor metaphorical. He wasn't grinning. He meant it.

Here was a man who effortlessly negotiated the porous, wafer-thin membrane separating Homo from the Other. "I am a puffin," he repeated softly yet firmly, "from my ancestral tree and in blood." Still alive. Neither museum artifact nor ethnographic data point. Standing before me. The flawless sheet of glass. Symmetrical, convergent consciousness: the world before the mind that thought in conversation with the earth, cracked, shattered into the mind that now observed the earth. *Cogito ergo sum* distills it in purest, crystalline form.

Except when, like Thoreau, we stand before the huge and waiting consciousness that spans the octave of creation. *Puffin is, therefore I am,* he told me that wintry afternoon.

Contact! Contact! Who are we? implores the slight, gray-eyed man into the swirling mist. Jesus of Nazareth screamed the same question, alone in the crushingly silent amphitheater we call wilderness.

For Faulkner the question loomed in the figure of an old bear in a Mississippi wilderness. *"So I will have to see him,"* resolves the boy, Isaac McCaslin, "without dread or even hope. *I will have to look at him.*" What was "this boy . . . born knowing and fearing too maybe but without being afraid, that [he] could go ten miles on a compass because he wanted to look at a bear none of us had ever got near enough to put a bullet in and looked at the bear and came the ten miles back on the compass in the dark"?[8]

In the crashing tonnage of a sperm whale Herman Melville experienced "some colossal alien existence without which man himself would be incomplete."[9] "In *Moby Dick,* Ishmael first tries to get at the mystery of the white whale through the science of taxonomy. He heaps up fact after fact about whales. . . . The limits of this method are reached, however, and still Moby Dick remains sovereignly unknown, uncomprehended. . . . The quest, to be successful, must risk the observer's life."[10]

The garment carries its own undoing. The bit of thread exposed. The Word scrawled on the sheet of paper. *I am puffin.* The *Logos* that whispers, contrary and conspiratorial and millennial, down the corridors of *Cogito ergo sum* till the terrible unexpected moment we are seized by mountain jungle desert whirlwind *Cetus Ursus Loxodonta* when suddenly without warning it swells like destiny, thunderous overwhelming shattering. *Contact!*

—Calvin Luther Martin

Notes

1. Rainer Maria Rilke, "The Man Watching," in *Selected Poems of Rainer Maria Rilke,* trans. Robert Bly (New York: Harper and Row, 1981), 107.

2. Henry David Thoreau, *The Maine Woods* (New York: Harper and Row, 1864), 94–95.

3. Rilke, "Man Watching," 107.

4. C. G. Jung, *Memories, Dreams, Reflections,* ed. Aniela Jaffé (New York: Pantheon, 1961), 272–73.

5. I Corinthians 13:12.

6. Jung, *Memories, Dreams, Reflections,* 277.

7. Daniel Quinn, *Ishmael: A Novel* (New York: Bantam, 1992).

8. William Faulkner, "The Bear," in *Three Famous Short Novels* (New York: Random House, 1961), 198, 241.

9. Loren Eiseley, *Francis Bacon and the Modern Dilemma* (Freeport, N.Y.: Books for Libraries, 1962), 94.

10. Leslie E. Gerber and Margaret McFadden, *Loren Eiseley* (New York: Frederick Ungar, 1983), 113–14.

Prologue

People often ask how I got involved with animals. I can't answer that, but I can recall my first encounter with the cold hard stones of what Calvin Luther Martin calls Adam's Wall: the rigid barrier separating us from other species.

I was about eight or nine. My parents did not like zoos, circuses, or even pet stores. Their disapproval was less spoken than simply understood through the scarcity of visits to these places. And so it wasn't my own family, but that of a friend, who brought me to an animal park.

The animal park was probably a zoo, but there were performing elephants, bird displays, balloons, and noise. After purchasing our tickets and pushing through the turnstile, we wound along cage-lined paths, corridors of cockatoos, strolling peacocks, and sleeping lions, up to the ape exhibit. A tall, grizzled keeper stood to one side answering questions about the animals in the cages behind him. Two chimpanzees sat on a concrete floor arrayed with a few barren branches cemented into the foundation. Pieces of browned fruit lay scattered about, and a third chimpanzee swung back and forth from a metal pole, looking down with a fixed stare.

Attention focused on the largest of the apes, who sat back on his heels, with knees up, facing the bars. Every so often, the keeper would call up a visitor so that he or she was standing in front of the big chimpanzee. At this point the keeper would chant: "Mirror, mirror on the wall, who is the fairest of them all?" The chimpanzee would then flail his arms and hands and break into a grin—the idea being that he was supposed to imitate the person standing in front of him. The crowd would laugh and giggle. As suddenly as it appeared, the chimp's disfiguring grimace would melt away and he would resume a motionless pose and blank countenance. I say "he" because that is how the keeper referred to him.

When we came close to the exhibit, my school friend ran up to the barred enclosure to be next in line. On cue, after the keeper made his incantation, the chimpanzee screwed up his face and waved his arms. There were squeals of delight and laughter all round, and then I was pushed forward to have my try.

Public attention is not something I enjoy, and even at that age, I sensed what I would now call the grotesquery in this entire display. But peer pressure and the protective arm of my friend's father propelled me straight to the cage facing the chimpanzee. By the time I got there, most of the crowd had drifted away. We were pretty much alone in the quiet, with faint bird calls and muted roars in the background, when the keeper began his chant "Mirror, mirror on the wall." I braced for what would appear, loathing every second and willing the ordeal to pass. But nothing happened. The chimpanzee sat motionless, face frozen in a passive stare. The keeper repeated the mantra. Again, nothing.

After three or four attempts, even my friends got bored and started to walk away, distracted by the gorilla or giraffe up ahead. "Well, I guess he just can't make a face as pretty as yours," the keeper said, obviously trying to make up for what he thought might be hurt feelings. "Why don't you go join the others, see the rest of the animals, and come back later?" he cajoled. I was more than happy to leave, and turned to walk away when suddenly a hand grabbed my arm. It was the chimpanzee. I turned to see him looking at me, and he made, as much as I could tell, a beautiful smile, his eyes soft. It lasted only a moment because the keeper started yelling and the chimpanzee's hand quickly withdrew. But in the split second that our gazes held, we had shared an understanding and stepped out of the Fellini world of the zoo. Despite the bars separating us, I felt closer to him than to any of my own species pressed alongside. Whether by chance or fate, my failure to play the game—the inability to evoke, or provoke, the chimpanzee's grimace—stopped my cultural progress, what the philosopher Paul Shephard calls the "lifelong work of differentiating" imposed by the modern world as we age from child to adult.

The revelation passed in the cavalcade of childhood adventures. No doubt when the keeper's yells broke the spell, I turned away and ran over to the giraffes to giggle and gawk with my friends. I forgot the chimpanzee, and it wasn't until years later that the memory came back like some mirage. In the interim, I became a scientist and plunged into the most human of pursuits—the disembodied mind. Life was divided into two halves, profession and family. Unmistakably, that moment served as a kind of pea under the intellectual mattress of, if not a princess, a restless scholar. The experience agitated a deep dissatisfaction and a restlessness that eventually led to the study of psychological trauma in elephants.

I came to the topic of elephant breakdown while visiting South Africa on invitation with other scientists to study lion reintroduction. During the 1990s, when apartheid was being dismantled, ecotourism began to rebuild. When it became clear that all-white rule in South Africa was ending, international travelers were ready to come back, and tourism again became a major and viable industry. Even earlier, parks and private reserves had been busy restocking with the Big Five (lions, Cape buffalo, leopards, elephants, and rhinoceroses) in anticipation of visitors hungry for South Africa's stupendous sights, long denied because of political tensions.

South Africa's wildlife reintroduction offered scientists a unique experiment to learn about lion population recovery and behavior, but there was another focus of inquiry. There were concerns about the consequences that this rather haphazard shuffling of genealogies and geographies might portend for long-term conservation success. Our first stop was Pilanesberg National Park, where I first learned about the elephants who have become so tragically intriguing.

Pilanesberg is located about two hours outside of Johannesburg, a small but marvelous landscape marbled with savannah hills. Like most other areas in South Africa, Pilanesberg had lost much of its wildlife, including its elephants, to hunting and poaching. It was therefore one of the reserves participating in wildlife translocation projects occurring throughout the country. Most of the elephants came from South Africa's own Kruger National Park, the flagship park of the country and, by some estimates, all of Africa. Kruger has not only tremendous faunal diversity but also a density of wildlife, including elephants, reminiscent of earlier times, before their decimation throughout Africa. However, their numbers were far from untouched. Park personnel controlled elephant numbers with routine culls, and the effects were significant. Young elephants orphaned from these operations formed the seed populations for diverse reserves to which they were moved to reestablish herds in places like Pilanesberg National Park.

As we bumped along the rutted road in search of lions, the park ranger began to recount the strange saga of the elephants. Park personnel and tourists were starting to find dead rhinoceroses, mysteriously gored. The numbers were growing, and even odder, the suspects were young male elephants. Stranger yet, the young bulls had been uncharacteristically harassing older female elephants; passing tourists had even caught one on film copulating with a rhinoceros.

My initial thought was that the elephants sounded a lot less strange than what I had learned earlier that day. Two former circus elephants had been reintroduced to Pilanesberg. An American elephant trainer, Randall Moore, struck by the plight of his longtime circus comrades, had saved enough money to send the two females back to their homeland. Years later, when he was able to come visit the park himself, he was reunited with his old friends. According to the ranger who accompanied Moore on his visit, the two elephants charged the jeep. (The ranger, of robust physique and experienced in the bush, confided that very little scared him except elephants.) He had been about to dash to safety when the two charging elephants suddenly stopped and sat back on their haunches, each raising one front leg. They recognized their long-lost trainer and were greeting him as in the old circus days.

We made our way back after a full day looking for lions. After dining early and exchanging observations from the day, we retired to rest. A few hours later, we rose. The air was chilly despite the day's heat. We put on jackets and jumpers, packed thermoses filled with hot tea, and jumped into the Land Rovers to head out into the tangled hills in search of two male lions. The air vibrated with insects and animal calls, and waves of unfamiliar scents—spicy, tangy, sweet—wafted by.

The park ranger had an uncanny ability to locate lions. Radio collars often helped, but the uneven hilly terrain challenged even this technology. We eventually came upon a small clearing, and after a few mugs of tea, two maned males appeared. I don't think that I will ever be able to gaze at a lion without awe and a butterfly fluttering in my stomach. Lions have a way of making the ego shrink and bringing out humility.

These two, however, sadly brought a sense of wrenching tragedy. Only a short while before they had made up a magnificent, intimidating threesome who daunted all challengers. Young males seeking mates and territory could not make inroads with this formidable trio and so were forced to distant park confines, occasionally even escaping through the fence. Lion extinction was so complete in the area that the sight of a stray male caused considerable consternation in surrounding communities. Gone were the days when villagers would accommodate lion habits by corralling their stock at night and avoiding nightly walks. Africa had forgotten how to live with its native wildlife.

Mindful of political sensibilities, the park rangers decided to stem the problem at its source—the alliance of the older males—and tracked and shot one of the three. The problem of roving lions stopped, but the two remaining lions were, in the words of the ranger, broken. The howls and roars of grief, he said, were heartbreaking. The two males stayed together, but their health deteriorated and they retreated into the bush to scrounge for food where they could. I could see that their coats were poor, and that they were thin and had an air of defeat. The ranger said that they were fully capable of retaining their former dominance, but it was as if they had lost their will with the death of their comrade.

The next day we left to visit other reserves. I left Pilanesberg—but it had not left me. The Pilanesberg story was reprised over and over in other parks that I visited. Perhaps few tales were as poignant as that of the two maned lost souls, but other animals faced similar problems as a result of human manipulations. At another park, reintroduced lions were mysteriously dying one by one. It was thought that genetic bottlenecking and inbreeding might be causing a loss of resilience and immunological vulnerability.

As the indescribably stunning landscapes of Africa continued to unfold, I began to grasp that the wildlife crisis in Africa and sadly elsewhere, including North America, was more than a matter of numbers. Humans were breaking the backs of animal culture and society with violence, even while seeking to conserve precious wildlife. In many instances, measures to repatriate and repopulate have exacerbated problems. I went home changed. Then, driven almost as mysteriously as the attacks on the rhinos, my research on elephants began.

This has not been an easy book to write for several reasons. The American anthropologist Nancy Schepfer-Hughes writes in "The Politics of Remorse": "I am taking on the politically and morally ambiguous task of telling the story of political violence and recovery in South Africa." Her moral challenge lay in recounting and creating what she calls a psychological anthropology "through the experiences, narratives, and points of view of a small number of white South Africans, drawn from across the social class and political spectrum." The challenge here is similar. It is perhaps even more dubious because it treads across so many conventional boundaries and in so doing is vulnerable to presumption. It trespasses into the history, lands, and lives of others and across species' lines, and in so doing, dares to presume an understanding of individuals who have experienced very different things and some who seem so very different. Furthermore, elephants' stories not only cross well-protected borders, they also dissolve them.

Elephant psychology is discussed here in much the same way as the profession approaches the human psyche. This has been possible only recently. The rate of new ethological discoveries concerning animal consciousness, self-awareness, empathy, culture, and so forth is breathtaking. Once a high scorer on the giggle-factor scale, the idea of an animal psyche has become acceptable (despite the reluctance of some) even in scientific research. These developments show how much our views of animals, and of ourselves, are changing. And as with most change, with it comes a certain challenge.

Then there is the issue of writing a book. Using words excludes participation to all but humans, and even certain humans, at that. Aside from its exclusionary nature, the written word has other limitations. A book is like a Polaroid snapshot—it freezes a life in the frame of two covers. The here and now become there and then. The elephant story is not fixed: elephant and human are ever evolving and contradict static representations. I have also had a sort of niggling unease with the implicit objectification that reportage entails—of committing yet another possessive trespass by modern humans and violating the rights and privacy of other species.

But the real obstacle in writing this book has been more personal. I have felt wedged between two worlds and struggled to bridge the chasm between the collectivity of science and personal nature of suffering—between my role as objective scientist and the subjective experience of a living, feeling, sentient member of the animal world. On one hand, colleagues advised, "Don't come across like an advocate or you will lose your credibility." Others warned, "Be careful that it's not too much of a downer or people won't read it." Yet others cautioned, "Whatever you do, don't sound like an academic."

Consequently, I started to write in a somewhat detached way, trying to salt painful facts with entertaining anecdotes, but I quickly became uninterested, the prose flat. I tried other voices, other ways, much like Goldilocks testing out chairs and finding something amiss with each. Finally, I ended up just writing about what I had seen and learned in the process of trying to understand what has been happening to elephants. At moments when stuck or struggling over a passage or concept, I would stop and ask: "What would you say if an elephant were standing right in front of you listening? What would you say that was honest and would not make you feel ashamed? Or sound too convoluted to be of any use?" Usually, I got an answer. Nonetheless, I felt shame at times, not so much from what I wrote but from what the words reveal: the brutality and horror that my species, my culture, I, have inflicted on animals.

Bringing words to page has been a soul-baring experience. The study of elephant trauma and the recognition of animal suffering in scientific theory and language no longer permit mental refuge from the devastation of wildlife genocide. Traumatology has awakened a new anguish by illuminating the radical contrast between the affectless language of science and the terrifying reality of environmental collapse embodied in the fall of a dying elephant, a drowning polar bear, and the silenced buffalo shot by government soldiers. Statistics describe what is happening less accurately than do images of screaming infant elephants tethered to their dead mothers after a cull, the vacant rocking of a bull elephant in a concrete zoo, the terror in the eyes of an elephant being "broken"—beaten with chains and rods to be made fit for service. These are descriptions that science typically avoids except through the disarming medium of numbers and theory.

As a teenager, I read *Night,* Elie Wiesel's testimony of his experiences in the Auschwitz and Buchenwald camps. The book still sits on a shelf, having occupied numerous other shelves and floors from college to graduate schools, squeezed between physics books on metal bookcases in basement cubicles, shoulder to shoulder with other old friends in the warmth of home, and in diverse weathered suitcases and backpacks as it accompanied me on travels in Denmark, China, Kenya, South Africa, Chile, and everywhere in between. The book became a sort of talisman, an unconscious conscience to safeguard any attempt at flight from honesty.

Shortly after I began writing about elephants, a new edition was published, a fresh translation by Wiesel's wife, Marion. I bought it expressly to read what he had written decades later in his preface. I am tempted to quote every passage, for indeed there is not one phrase that does not burn into memory. However, I encourage the reader who is not familiar with Wiesel's book to read it, and I limit inclusion here to one passage because it pertains to a sense that I experienced in writing about elephants. "In retrospect, I must confess that I do not know, or no longer know, what I wanted to achieve with my words. I only know that without this testimony, my life as a writer—or my life, period—would not have become what it is: that of a witness who believes he has a moral obligation to try to prevent the enemy from enjoying one last victory by allowing his crimes to be erased from human memory."

The enemy of whom Wiesel speaks seems obvious, particularly from the perspective of decades' remove: Hitler and his accomplices, as well as those who later denied that the Holocaust ever happened. But the enemy about whom he writes is also the "enemy within," the potentiality of violence and cruelty in every individual, and the complicity of those who see, know, and yet remain silent.

The spirit and purpose of this book is similar: to try to see through the layers upon layers of assumptions and assuaging myths that have muted the suffering of elephants and allowed their continued slaughter even in the presence of knowledge and information. To see individual elephants and their cultures, and to witness, if not though their eyes, through, as the Fox from *Le Petit Prince* suggests, the heart. A way of seeing that meets elephants through the lens of a new paradigm, a trans-species way of knowing that we are "kin under the skin."

In the following pages, we step into the feet of these giant, mythical herbivores and grasp some of what elephant experience has been and is. Through the narratives of neuroscience, dramas of individual lives, and historic events in which elephants play a part, we get a glimpse into the pulsing soul of a great species and insights into our own. The written word as it appears here is used almost like nail polish remover: to dissolve the cultural veneer that has occluded a once open channel for dialogue across species.

In the closing passages of his preface to *Night,* Wiesel reflects on a question he is often asked, whether he knows the "response to Auschwitz." He writes: "I answer that not only do I not know it, but that I don't even know if a tragedy of this magnitude has a response. What I do know is that there is 'response' in responsibility. When we speak of this era of veil and darkness, so close and yet so distant, 'responsibility' is the key word."

On behalf of the Indian Nations, Vine Deloria Jr. demanded of white America, "We talk, you listen." I think that the elephants' plea is much the same. By listening, we are asked to ethically and practically challenge what has seemed indelible—the uniqueness and privilege of being human. We are asked to be a part of elephant suffering, to own what has happened and is happening to elephants as something that involves us as much as them. The elephant experience entreats us to come around the fire ring, to sit and listen while the embers burn and the clear night domes above, much like our ancestors did thousands of years ago. For in the elephant's tale is a story of our own past, and, poignantly, what can be our future. Even more important, it is not so much that elephants are *like* us. They *are* us, and we them. That is the lesson the chimpanzee instilled in me, and what I hope to pass on to you.

Acknowledgments

Many have been involved in the development of this book, some directly, some less so, but all have contributed immeasurably. Without the sterling heart and mind of Dame Daphne Sheldrick D.B.E., and her generosity, this work would not have been possible. I would also like to give appreciation to Carol Buckley and Pat Derby for sharing their experiences with me and for showing the world what a happy elephant in captivity really looks and sounds like. Suparna Ganguly, Sandra deRek, Catherine Doyle, Debbie Leahy, Suzanne Roy, Don Elroy, Dorothy Phillips, Mary Robinson, Melissa Groo, Lisa Kane, Nicole Paquette, Mary Robinson, Les Schobert, Deniz Bobol, Margaret Morin, and Penelope Wells gave magnanimously of their time to share details, garnered through diligence and dedication, of elephant experiences.

My thanks for the generous support of workers who have contributed so much to deepen understanding of all animals: Cynthia Moss, Keith Lindsay, Phil Ensley, Allan Schore, Delia Owens, Michele Pickover, Elke Riesterer, Melissa Groo, Mel Richardson, Evelyn Lawino Abe, Angela Sheldrick, Steve Best, Leslie Irvine, Betsy Swart, Marc Bekoff, Joyce Poole, Marion Garai, Fred Kurt, Robert Sapolsky, Iain and Oria Douglas-Hamilton, and Katy Payne. Keith Lindsay, Mark Owens, and Mike Cadman deserve special gratitude for their scholarship, dedication, and honesty in a vital cause and were ever timely and helpful as I made my way through the tangle of elephant politics, history, and science. Warm thanks goes to Harry Biggs and Rina Grant-Biggs for their marvelous contributions to science and issues pachyderm, their unwavering friendship, and their wonderful books.

It is certainly not uncommon for an author to have a sense of gratitude for the publisher, but in this case special thanks are in order. The dynamism, expertise, enthusiasm, and humor provided by my editor, Jean Thomson Black, were vital in helping to shape this book. Thank you, Jean. My manuscript editor, Dan Heaton, provided the tough love that renders the rough hewn into polished stone. Anne Borchardt made this process elegantly easy and her belief in this project is ever appreciated. It has been an honor to work with such professionals. Very special thanks to Amy Mayers for research and editing assistance. Many thanks to Deb Robinson, Cynthia Corbit, Alexis Soulios, Joe Calamia, Kate Johnson, and Julie Lockhart for editorial and research assistance and for providing on multiple occasions "le mot juste," and to Kent Garber for the many details with which he helped in the manuscript preparation. Like most authors, I was saved from literary myopia by colleagues, new and not so new, who provided essential perspectives during the preparation of the manuscript and gave generously hours of patient support, listening, disagreeing, and agreeing. My profound thanks go to Lori Marino, Peter Mudd, Toni Frohoff, John Gluck Jr., David Bjorklund, pattrice jones, Phoebe Greene Linden, Nina Pierpont, Martin Brune, Joseph Yenkowsky, Diana Reiss, Eileen McCarthy, Josh Plotkin, Karen Paolillo, Robert Jay Lifton, Robin Bjork, Joe Mitchell, Patty Finch, Debra Durham, Terri Jentz, Gloria Grow, Susan Griffin, Mira Tweti, Neva Yarkin, Laura Stille, Ruth Klaus, Celeste Wiser, Tim LaSalle, Susan Donohoe, Susan Konecny, Ginger Casto, Mary Watkins, Jerome Bernstein, Elizabeth Nelson, Art Glenberg, Ann Southcombe, Fariborz Rostami, Lee and Earl Showerman, and Ken Shapiro. Thanks also to Gary Weaver and the Veterans Administration. Cyril Christo and Marie Wilkinson deserve a big bow for bringing the beauty of elephants and other animals to those who would never see them otherwise and for making their marvelous art part of this book. Many thanks to Charles Siebert for bringing Emily Dickinson to the elephant world and back again. I would like to express deep appreciation for Ray Ryan, Henry Krystal, Lola Gonzalez, Renaldo McGirth, the Grimes family, and Deogratias Bagilishya for their generosity, integrity, and willingness to share their stories and insights with such grace. The time spent with each of them is truly cherished and will always be remembered. My gratitude goes, too, to the elephants themselves, as a culture and individually, including the three South African bulls, Echo, Jenny, Dunda, Billy, Girija Prasad, Peaches, Flora, Minnie, Shirley, Maggie, Pet, Benjamin, the Divas, Sissy, Teoha, Osh, Mary, Rose-Tu, Hansa, Delhi, Dunda, Ndume, Dika, Imenti, Edo, and Ned. Thanks also to the reptile clan: Garline, O. J., Ralph, Soila, T, B, and many others. I would like to offer an apology for not knowing whether your participation was willing, but I hope with all my heart that it was. I am so sorry for your suffering but know that your stories will live through time to stop future violence and pain and offer a different way of living than that you had to endure.

Perhaps no one spent as many hours with things psychological and philosophical than my bomba brilliant colleague and friend Deanne Bell; thank you. Vera Müller-Paisner provided unique insights on topics ranging from horses to heuristics to herring—thank you for and from mind and heart. Vine Deloria Jr. is ever an inspiration and a beacon of integrity for us all. This book is written in memory of him and of the bisons whose tracks once spanned North America and, like those of the elephant, have all but disappeared, but nonetheless will flourish again. Calvin Luther Martin has been a poet, friend, and mentor, always providing an unspoken reminder to keep the faith. My gratitude for your support and gifts of the heart and mind. O. Mein-Gans, my soulmate, who has stood by unwaveringly with love, never doubting, I can never thank you enough. Finally, to Joseph P. Bradshaw Jr., Vivian O. Bradshaw, Joseph P. Bradshaw III, Jeffrey Borchers, Carol Bradshaw, Chauncey and Dorris Goodrich, Barbara Dallas, Barbara Harrell, Carl Harrell, Paul Dallas, Donna Wilshire, Bruce Wilshire, Ty Bradshaw, "them," and S. Nake—you all know what I am thinking.

A Note on Terminology and Sources

This work draws from multiple disciplines and cultures. It is, therefore, informed by a diversity of terms that are used to describe the same or similar phenomena. To avoid excessive, iterative clarifying definitions that clutter the narrative, a certain linguistic convention has been adopted. Unless otherwise specified, *elephant* is used with the assumption that the characteristic being described generally holds for all elephant species and subspecies. *Captive* or *captivity* pertains to elephants in the closed confinement characteristic of zoos, circuses, and other settings of extremely limited space; the terms imply attendant psychological and physical restrictions. *In captivity* is used in lieu of *captive* to emphasize that it is the conditions, not the elephant, that differ behind bars or fences compared with animals who are free ranging. *Free-ranging* (occasionally referred to as *wild*) elephants live in conditions that most closely resemble, but do not match, those of precolonial, historic times. *Sanctuary* is a human-made environment created to rehabilitate animals who have suffered psychologically and physically under captivity.

In clinical studies, participant anonymity is protected. Researchers are required to procure each subject's consent or, where competency is uncertain, approval of a guardian or supervising physician. Protection holds even for deceased subjects. No consensus exists for parallel treatment of animals other than humans. I have used given names in lieu of maintaining anonymity in an effort to discontinue the elephants' objectification. In cases where there was a guardian at sanctuary, permission to write about an individual elephant was obtained.

Finally, a note regarding sources and references used in this work. It is the nature of the study that many draw from media reports as well as the peer-reviewed scientific literature. However, only media sources that were substantiated elsewhere or derived from public records were used; they are considered to constitute data that are supported in more formalized journal articles and texts.

1
The Existential Elephant

She is middle-aged, amiable, but known to keep to herself. Unlike others her age, she has no children and there are no men in her life. She has two neighbors who are contemporaries, Patty and Maxine, but the two have formed a fast friendship, leaving Happy something of a third wheel. According to records, Happy was born in 1971. As an infant, she was forcibly taken from her family in the Asian wild and shipped to the United States. After five years at a facility in West Palm Beach, on March 21, 1977, she was transferred to the Bronx Zoo, where she received the identification number 771057.[1] *For a short while she had a companion, Samuel R., a teenager almost young enough to be her offspring, but the youngster sickened and died. Happy has never experienced the traditional family life of her native culture, in which she would have spent her life with her mother, sisters, cousins, and aunts, caring*

for babies and mentoring teenagers. The Bronx Zoo announced that the elephant exhibit will close when one or two of the trio dies. Happy will probably live out her days alone in New York City, unless she can be sent to sanctuary.

Who is an elephant? A simple-sounding question, but one that launches us on a long and fascinating journey into unexplored terrain. The idea of an elephant being a "who," not a "what," is somewhat unfamiliar. Even questions of human identity are complex, and the riddle of the self has haunted humankind throughout history. The French philosopher Jean-Paul Sartre and his lifelong colleague Simone de Beauvoir sat for hours on end in smoky Parisian cafés debating the idea of existence, its meaning, or lack thereof. Today, philosophical reflections on the nature of the self are tackled in the language of neurons and synapses, but the essence of the search remains, and while few of us have written weighty tomes on the subject, most recall at least one time when we set out to find out who we really were.[2]

One of the ways we get a glimpse of this entity called the self is by gazing into a mirror. Copper was crafted in Egypt to reflect the human face as early as 2900 B.C.E., and mirrors of all sorts of shapes and sizes inhabit myths, legends, and fairy tales around the world.[3] Snow White and her wicked stepmother would hardly be the same without "mirror, mirror on the wall," and Alice could hardly have discovered the things she did without her looking glass. Mirrors are mysterious; they entreat the viewer into a world beyond the ordinary, something and somewhere that reveals what an outward gaze cannot, an inner identity and sense of "meness" that is more than skin-deep.

Musings on existential meaning and looking glasses may seem a far cry from the world of bloodied rhinoceroses and stalking elephants, but they relate intimately to the tale. Elephants are among the few nonhuman species known to recognize themselves in mirrors. This, according to scientists, demonstrates self-awareness. Gordon Gallup is a psychologist who is credited for inventing the "mirror test," a method of investigation intended as a way to probe the mind's inner world—the means by which the elephant self was formally discovered to exist.[4]

The Mirror Self-Recognition (MSR) test identifies critical levels of cognitive development, the ability to engage in abstract psychological levels of knowing that relate to the theory of mind (TOM), indicating that one is not only conscious but self-conscious. The assumption behind the test is that only a subject who has a sense of self, an awareness of her existence as a unique being and who acts as an instrument of her own fate, is able to recognize her form in a mirror's reflection: to know what she looks like on the "outside." Though philosophers and scientists have had a difficult time agreeing on what lies behind a sense of self, and even its definition, the MSR test purports to objectively measure something as subjective as the interior felt sense of self-awareness.

Self-recognition typically develops in humans between the ages of eighteen and twenty-four months; until recently, it was believed to be absent in animals and assumed to distinguish humans from all other species. But in recent years this exclusive club has been joined by chimpanzees, orangutans, dolphins, and magpies, and it was Happy, the middle-aged Asian elephant living in the Bronx Zoo, who carried home the intellectual gold medal to qualify the pachyderm for membership.[5]

Despite its popularity, the mirror test is a subject of some debate.[6] Criticism comes from both those who doubt the equivalence of animals with humans and those who do not. Skeptics maintain that the test is arbitrary and that each species may respond differently based on variations in species-specific and individual characteristics and motivations unrelated to any intrinsic ability of self-awareness. Measures and meaning of self may not hold universally given species' diversity. Other argue that there may be systemic species differences that relate to patterns of evolutionary pathways. Neither do responses appear consistent within species. Some researchers report self-recognition in gorillas, whereas others have found the opposite, and not all chimpanzees test successfully.[7] So when scientists decided to investigate the elephant self at the Bronx Zoo, the experiment was meticulously designed to meet specific protocols, then documented and filmed.

More is involved than merely hanging up a mirror and watching elephants. The subjects are stepped through a sequence of stages to ensure that their behavior really reflects what the test is meant to demonstrate.

Happy, Maxine, and Patty were introduced to the mirror only after it had been kept covered for several days in the outdoor area, in order to contrast behavior related to the appearance of something new in the environment—no small event for an animal in confinement—with that specific to an awareness of the mirror's reflecting quality.

Zoo life is routine. Residents follow a schedule shaped to human schedules and customs. Feeding, bathing, veterinary check-ups, training, and exhibition times are regimented, and living quarters are spare compared with life in the wild. Exhibits are generally concrete, with bars and barriers, and not infrequently animals are bereft of companionship. Even if there are other conspecifics (others of their own species) with whom to interact and socialize, relationships are often temporary, as animals die prematurely, fall ill, or are traded between institutions.

Elephants have evolved with the natural rhythms of weather, landscapes, variations in food and water, and the cycles and drama of birth and death. Free-ranging elephant life is rich with relations, color, flora, fauna, and freedom, and days are spent on the move, with nearly constant social contact while foraging for grasses, bark, leaves, twigs, and roots of diverse botanical species. The family group covers tens of miles daily, roaming across terrain that demands physical strength and mental competence.

Zoos have looked for ways to make life more bearable and interesting for their residents. Plastic balls, barrels, tree branches, and treats are intended to "provide environments of greater physical, temporal, and social complexity that [afford animals] more of the behavioral opportunities found in the wild."[8] Such objects are meant to break the monotony of confinement and compensate for what captivity lacks and what elephants need.

It is no wonder, then, that when a full-length mirror is brought into the stalls of the Bronx Zoo, the sophisticated but underchallenged minds and senses of Happy, Patty, and Maxine might just be reacting to something new. The subjects testing for MSR work through four stages, each of which the subject must pass to demonstrate self-recognition. The first stage is referred to as the "social response": when an individual sees her reflection, she may initially think it is someone else and react defensively by roaring, running away, or trying to interact in some fashion. But the Bronx Zoo elephants did none of these things. Instead, they skipped ahead to the second stage, called "investigative behavior," using their trunks to probe over and behind the mirror and even tried to climb over the wall to see what was behind it. Happy, on the other hand, did not investigate behind the mirror.

After only three or four days, all progressed to the third stage of the experiment, which determines whether the elephants will investigate themselves—a test of self-consciousness. They carried their food over to eat in front of the mirror, using their trunks to touch parts of their bodies. For example, Maxine put her trunk tip first into her mouth and at another time pulled her ear slowly toward the mirror. These behaviors were interpreted as evidence of mirror use. But Happy excelled all the others when, on the first day, she passed the fourth and final stage, the "mark" test, which reveals self-directed behavior or actual signs of recognition of herself in the mirror.

A white cross was painted on one side of the individual's forehead such that it was not visible without the use of a reflective surface. A "sham" invisible cross was also painted but on the opposite side of the forehead, to eliminate the possibility that the white cross could be detected through smell or feel. Happy touched the white cross with her trunk while standing in front of the mirror, but she lost interest after the first day. Even though she continued to approach and explore herself using the mirror, Happy never touched the mark on her forehead again, and the second day did not even visit the mirror. Maxine and Patty showed no interest at all, technically failing the mark test. However, together, the evidence was sufficient to qualify a pass on the MSR test and demonstrate elephants' capacity for self-awareness.

All this may seem like a complex scheme just to show that an elephant can use a mirror, but the study provides ethological evidence that parallels findings from neuroscience: a trans-species model that simultaneously describes human and other animal brains and behavior.[9] Dr. Lori Marino, a neuroscientist in the neuroscience and behavioral biology program at Emory University, has made the point that the neurobiology and behavior involved in mirror recognition are strikingly similar across species. Whereas Happy and her compatriots brought hay to the mirror and ate in front of it, dolphins sometimes bring their toys. As in humans, the capacity for self-recognition appears to be developmentally linked. Young chimpanzees exhibit self-awareness over a range of ages, some as early as two and a half years old. Dolphins just over a year old have shown a fascination for their reflected images.[10]

Such similarities are not surprising, and ethologists and others have accumulated an impressive list of examples. All vertebrates (and, some would argue, invertebrates) have a common suite of brain structures and mechanisms that underlie complex processes of cognition and affect, including areas of the brain that involve self-awareness, decision making, and planning.[11] Key cortical and limbic structures in the brain are conserved evolutionarily across species, as are areas in the brain responsible for processing and controlling emotional and social information associated with specific psychophysiological and behavioral traits. We "social brained" animals, whose circuitry evolved to interact and deal with our conspecifics in complex ways, have developed shared structures and functions in the brain that engender similar capacities to think, feel, and behave. Maternal behavior, the ability to recognize faces and expressions, play behavior, sexual behavior, fear, aggression, and emotions and feelings are but a few items on a very long list shared across the animal kingdom. In this new trans-species world, dolphins have culture, crows use tools, sheep empathize, and snakes play. Fish subjected to electrical shocks retreat into dissociative rocking as a means of coping with pain, much as human victims of torture might.[12]

The fact that we can be so different on the outside yet otherwise so similar is both fascinating and eye-opening. We have learned something not only about elephants but, as Gordon Gallup predicted, also about ourselves. "The history of science . . . can be viewed . . . as having brought about gradual changes in man's conception of man, and with such changes man may eventually have to relinquish, or at least temper, his claim to special status."[13] By having a sense of self, elephants and other animals erase a significant delineation of human uniqueness.

Historically, the sense of a personal self has defined much of what it means to be human. But not all cultures have such strongly held assumptions. Many American Indian cultures regard other animals as fellow tribes—the Wolf People, the Buffalo People, and the Eagle People—each with its own language, customs, law, and land. In contrast, the concept of human uniqueness has been an organizing nucleus of Western, Euro-American culture and informs the basic rules of everyday life—social rules and laws, eating habits, cultural practices, and science. What may seem an immutable assumption today actually represents the culmination of a long intellectual and political evolution.

The separation of humans from other animals dates back to the Stoics, who defined *hegemonikon*—the highest component of the soul—as the personalized, private imagination. Along with consciousness and awareness, hegemonikon brought an individualistic definition of spirituality to the communion between God and humans. Concepts from the Berber Saint Augustine of Hippo deepened the split by creating an image of self as inner space apart from the outer world. Later, during the Reformation, the self became inextricably identified with the religious and political identity and autonomy that have become the collective agenda of liberalism, capitalism, and the broader agenda of human dominion over all other creatures.[14] The portrayal of a singular, bound human self relates to the overall pattern of splitting that dualism has created: humans/nature, mind/matter, and so forth.

Consequently, when Happy and her predecessors stepped across the divide between those who do possess self and those who don't, more than a few sacred and unwitting bovines of Western civilization were challenged.

For instance, the question of whether an animal has a self turns attention to the culturally accepted premise of keeping Happy and other animals confined in zoos. In the absence of a sense of self, keeping elephants in zoos seems less inhumane than human captivity. But once science demonstrates that animals have joined humans in this one crucial respect, doubts are raised about the legitimacy of institutionalized captivity, and about owning animals, using them in experiments, and consuming them as food. Self-recognition is one of the more formidable and controversial scientific salvos against the long-held assumption that animals are fundamentally different from humans.

But the burden of proof does not lie on the shoulders of the mirror test alone. The decades of ethological observations collected conform to criteria identified by the psychologist William James to describe the core self and have laid the foundation for extending the concept across species.[15] James identified four abstract features of the "I" that relate to actual lived experience and felt action and give us each a subjective sense of self.[16]

First, a sense of self involves agency, the knowledge of one's own behavior and movements that acts on the world. Second, this sense of agency feels coherent and embodied. According to developmental psychology, self-knowing relates to the moment when an infant first realizes that Mother or Father is "not me," thereby initiating a process of awareness that evolves over time. The incipient and growing sense of agency relates to the understanding that when "I" throw the ball at Billy, it is Billy who feels the pain, not the "I" who caused it, and who may subsequently live to regret my impulsive action when I become the embodied target of Billy's wrathful agency after he recovers.

Given James's first two definitions, it is hard to imagine that any animal would not qualify. A sense of agency is vital for survival: how do "I," not Mum or Dad or Billy, know what and how to eat, interact with another, or decide which trail to take in the woods? Self-capacity is consistent with current behavioral models that describe elephants, humans, and fish alike. Historically, animals were envisioned as functioning much like organic, genetically programmed robots, their behaviors largely or exclusively instinctual, with predetermined responses awaiting appropriate stimuli. In this view, agency was in no way required to prompt neurons to fire and muscles to contract as the animal learned how to stalk, pounce, and gobble up a grazing gazelle, how to decide who is friend and foe, or whether to come running at the sound of the cry of one's infant. Elephants offer a superb illustration of and necessity for Gallup's bidirectional consciousness.

It is by dint of the exigencies of social life that elephants know and remember "who's who," self included. Not only do they recognize each other out of hundreds of other elephants, but each understands who and what he is relative to this complex hierarchical web of relationships. As part of a social network that once stretched across Africa, elephants adapted to a multitiered social hierarchy that demands the ability to recognize and distinguish among themselves. Much like avid fans of soap operas like *Dallas* and *All My Children,* elephants, and in particular the matriarch, must be able to recall myriad relational connections and intrigues that make up the fabric of everyday social life upon which survival depends. The matriarch is vested with social and ecological knowledge garnered over the years that guides her decisions, life-and-death choices that keep members of the group safe and healthy.[17]

Elephants' ability to discriminate through smell, vocalizations, touch, and sight extends beyond kith and kin. Elephants have shown that they can also distinguish among humans. Lucy Bates and her coresearchers discovered that elephants react differently when presented with red clothing worn by the Maasai people (who kill elephants) and with clothing worn by the agricultural Kamba tribe (who do not).[18] Shown Maasai clothing, elephants became startled and ran away, whereas they displayed little reaction to Kamba clothing. The elephants were also more perturbed when presented with red as opposed to white clothing, an effect that researchers inferred was related to the Maasai's frequent use of red. As we shall see later, Dame Daphne Sheldrick, founder of the David Sheldrick Wildlife Trust outside Nairobi, Kenya, makes the same point: elephants can distinguish between who has hurt them and who has helped them.

Elephants have been discovered doing other "humanlike" things. They are vocal learners, similar to parrots. For example, they have been heard mimicking the sound of trucks and lawnmowers. Nor are they strangers to tools. Asian elephants commonly use their trunks with remarkable dexterity to position and agitate a well-placed stick to scratch a hard-to-reach spot on their challenging bulk or to shoo an irritating fly.[19]

Cognitive acrobatics, memory feats, and intricate social interactions all point to a degree of individuality that contradicts past claims that elephants, and animals in general, are little more than wind-up bundles of instincts. When you see a stick grasped by a coiled trunk, navigated through space, and delicately directed to a specific area amid the gray sea of the vast pachyderm frame, then lazily pulled back and forth to relieve an itch, it's hard not to recognize ourselves and recall that same pleasure.

All this accounts for why—despite how different we may appear—when we look at an elephant, we recognize so much of ourselves. Neuroscientists have found another explanation for why we can feel and anticipate what an elephant is experiencing. The answer is mirror neurons, tiny brain cells wired for empathy that fire up at the sight of an elephant scratching, or that make us salivate while watching someone eating an ice cream cone on a hot summer day. The discovery of mirror neurons has articulated a much-needed bridge that can span the gap between cognition and biology by providing a neural mechanism for the psychological phenomenon of empathy.[20]

What is most intriguing here is not just the discovery of a cross-species physiological basis for emotional states such as empathy, but also the implications about who the self really might be, and where the boundaries between you and me and between Happy and the zoo visitor start and end. The sensate biological connection between two individuals that is activated by mirror neurons suggests that perhaps our sense of self, that ineluctable essence, derives from something beyond the individual, a realm where "I" is more than one. This brings us back to William James's checklist and his last two definitions of self.

The third feature James associated with a sense of self was the capacity to feel and show emotions, particularly when there is a clear demarcation between the sensor and the sensed—that is, between one person and another person. Studies on elephants in the African and Asian wilds and those in captivity indicate that in addition to complex cognitive skills, elephants feel and express a range of emotions. Elephants are well known for showing wrenching grief when a loved one dies or somehow departs forever, as was probably the case with Happy when she was captured and taken away from her mother. Perhaps more than any other quality, elephants show an understanding of death with their displays of grief and repeated visits to the bones of relatives.[21]

George Adamson, who with his wife, Joy, inspired the film *Born Free,* is one of many who have seen elephants burying their victims. In his memoirs, Adamson recounts how he was led to an area where the earth was furrowed and was littered with bits of rags and bones. This marked where an elephant had killed and then buried a man. Even after the burial, the elephant would return and visit the grave each afternoon to plow up the ground around the grave with his tusks.

On another occasion, Gobus, a Turkana game scout for Adamson, related that his half-blind mother, after finding herself lost in the woods, had decided to settle down for the night under a big tree rather than risk an encounter with a nocturnal predator. Suddenly in the night, she awoke to find that she was surrounded by elephants. One by one, each proceeded to drape her with branches and leaves until she was completely covered. When her son discovered her in the morning, they both realized that the elephants had presumed her dead and were burying her as they do with humans whom they have killed. Others have also observed elephants covering their own dead with branches and dirt.[22]

Like us, elephants perceive the emotions and anticipation of others and extend their empathy to companions who are injured or ailing. Grieving and mourning rituals make up an integral part of elephant culture. A mother may grieve over her dead child for days after his death, alternately trying to revive the baby and caressing and touching the corpse. Cynthia Moss and Joyce Poole have observed a mother risking her own life for a week to grieve over her stillborn child.

The death of a matriarch is particularly difficult for the community. Senior females form the pillars of elephant communities, and when they die, the entire herd is affected. A matriarch's death means not only the loss of a loved one but also a loss of cultural and environmental knowledge. An elephant matriarch's uncanny memory and her ability to process complex social and ecological information to successfully guide her family to food and safety can be traced back to her neuroanatomy: magnetic resonance imaging reveals that an elephant possesses an extremely large and convoluted hippocampus, the brain structure most responsible for mediating long-term social memory.[23] But any death in the family is significant. When Emily, in the "EB" group studied for many years by Cynthia Moss, died in 1989, it was "the greatest trauma in this family's life" since 1973, when Moss had first met them.[24] The deaths of calves are distressing for their mothers, but the death of an adult female disrupts the whole family.[25]

After Emily's death, the group performed mourning rituals. Later, when time had dissolved the last vestiges of her massive flesh, her whitened bones lay spare, but not forgotten. For years the aftershocks of Emily's passing could be observed as the group visited her bones.

> The three animals stopped and cautiously reached their trunks out. They stepped closer and very gently began to touch the remains with the tips of their trunks, first light taps, smelling and feeling, then strokes around and along the larger bones. Eudora and Elspeth, Emily's daughter and granddaughter, pushed through and began to examine the bones, and soon after Echo and her two daughters arrived. All the elephants were now quiet and there was a palpable tension among them. Eudora concentrated on Emily's skull, caressing the smooth cranium and slipping her trunk into the hollows in the skull. Echo was feeling the lower jaw, running her trunk along the teeth—the area used in greeting when elephants place their trunks in each other's mouth. The younger animals were picking up the smaller bones and placing them in their mouths, before dropping them again. . . . Several years before, I had also seen the EBs start to bury the carcass of a young female from another family who had died of natural causes.[26]

Genetic studies now confirm ethological observations: researchers have found that it may take up to twenty years for a family to rebuild after suffering the devastation of loss and fragmentation from poaching and other forms of mass killings.[27]

The same emotions follow elephants into captivity. Pat Derby is the director and founder of the California sanctuary Performing Animal Welfare Society (PAWS) established in 1984, where rescued African and

Asian elephants and other "refugee" animals from zoos, circuses, and the entertainment industry are given a home. She recounts another example of deep bonds, when Tami, the companion for forty years of an elephant named Annie, died:

> I have never seen such grief. We thought we were going to lose Annie. She just became passive, and when she looked up, it was as if she looked right through us. The only thing that seemed to give her pleasure was to float in the Jacuzzi that we had built for Tami. After Tami died, Annie refused to lie down to sleep at night. It was only in the Jacuzzi that she would be able to get some rest. She would walk in and then float and nod off and relax. For seven months she lived that way. Then suddenly, one day, it was like she had turned off a faucet inside. We were leading her up the hill when she suddenly stopped halfway and looked at us with the funniest expression. Then she turned around and went down to the lake to be with Minnie and Rebecca. She was ready to be with others again. It's as if she had to have this period of mourning for Tami, and then that was it. You can tell she hasn't forgotten Tami. Even though she gets along with the other elephants, she's still off into herself. You know her relationship with Tami was so special that there is no one else who can take her place. She'll be devoted to Tami's memory the rest of her life.[28]

Outside the confines of captivity, bonds are just as strong, and an entire group may pace its progress to match what an ailing member can manage. In Kenya, when a young African elephant, Ely, was born crippled by poorly articulated carpal joints, his mother and older sister stayed with him, assisting and prodding him along. Martyn Colbeck recalls that "the threesome headed toward us through the picturesque palms of Ol Tukai Orok. As the two older elephants walked, they continually turned to look back at the calf that was shuffling along behind. Every few feet they stopped and waited for him to catch up before moving on. Their progress was very slow, but they showed no impatience. It was a poignant sight and highlighted the incredibly caring nature of these animals."[29] Eventually Ely was able to walk unassisted.

Elephant emotions and empathy may be novel to Western science and cultures, but in places where people have lived side by side with elephants for millennia, such skepticism is trumped by experience. When an elephant was killed recently in Jharkhand, India, there was concern, and one official observed, "When an elephant dies, others keep coming to that spot for three to four days. We have deployed our team with all necessary resources as a precautionary measure."[30] The Indian official's worry was rightly placed. With humans after a violent death, emotions may run high and relations between groups can be tense. Similarly, at an elephant wake, as with human memorials, family members will be distraught over the passing of their comrade, and any misunderstanding can spark further violence.

This brings us to William James's fourth and final quality defining a sense of self: possession of one's own sequence of experiences, a history and sense of continuity. John Locke asserted that continuity of self—meaning a feeling that persists over time of being who you are—derives from a concatenation of memories. As we have seen, memory is a marvelous adaptation to a social and biological environment that requires intimate knowledge about one's friends and enemies.

In zoos and circuses, elephants are known for what has been called their "retaliatory cunning," a calculated, directed attack on someone who has harmed them in the past. In more formal psychological terminology, retaliatory cunning relates to "autobiographical memory," long-term recollections of significant personal experiences. It is self-knowledge that involves engagement with the external world, memory of self and of experiences with people and places, and the ability to wait, plan, and seize a propitious fleeting circumstance to redress past mistreatment or preempt it in the future.

After experiencing injury in the course of training for circus performance and other types of entertainment, elephants sometimes seem to use patience as part of a strategy to extract vengeance or to prevent further injury by the same individual. Numerous trainers, veterinarians, and keepers have given accounts of elephants who remembered the agents of their suffering and waited for years to exploit an opportune moment. The results are sometimes fatal for both humans and elephants.

The veteran "elephant tramp" George "Slim" Lewis recounts how a male circus elephant, Black Diamond, struck out at a former keeper who had been particularly abusive in the past but whom the elephant had not seen for many years. In the process, Black Diamond knocked down and fatally gored a tourist standing nearby who was trying to pet the "rogue" elephant. In almost all such incidents, the attacking elephant is summarily disposed of.

Hours after the incident, three of Black Diamond's legs were chained to three other elephants, and he was led through streets lined by thousands of shouting, vengeful Texan spectators. Upon arriving at the circus grounds, the condemned bull was tied to his block and his leg chains were transferred to trees on either side. He was then given several bags of peanuts, some of which were poisoned. According to Lewis, Black Diamond tentatively examined them and slowly and carefully chose only the peanuts that were not poisoned. After this unsuccessful execution, he was offered poisoned oranges, which he refused. Finally, more than 170 rounds were shot into him, and Black Diamond "died without making a sound or fighting his chains." His bullet-ridden head was mounted with artificial tusks and displayed in Houston.[31]

In light of what we are beginning to discover about how closely elephants resemble us emotionally and otherwise, Black Diamond's memory-inspired aggression should not be confused with the rampages of young South African elephant bulls who killed more than one hundred rhinoceroses and who, as we shall see, play such a pivotal role in our exploration of the elephant mind and psyche. These individuals were eventually diagnosed with posttraumatic stress disorder (PTSD), and while Black Diamond might also have been a prime candidate for a similar diagnosis given his experience of violence in the circus, their histories differ.[32] An attorney who attributes his client's angry outburst to war experiences makes a different case than she would in defending Romeo for slaying Juliet's cousin Tybalt in a grieving rage after Tybalt has killed his beloved friend Mercutio. To understand elephant experience, it is necessary to continue unraveling elephant psychological mysteries—to understand individual differences in elephants in the same way we try with humans. Something that is now scientifically possible.

C. G. Jung once wrote that humans remain a mystery to themselves because they are unique, lacking someone or something against which comparisons can be made. This argument can no longer be made. The practice of using animals as human surrogates for probing into the human mind and human behavior implicitly acknowledges cross-species similarities, but somehow, though sharing the attributes that privilege humans, animals have been denied psyche and rights.

Today, the seemingly impermeable species barrier has eroded, similarities outweigh differences, and a theoretical and perceptual fusion has taken place. Human psychology and animal behavior are brought together in the creation of a trans-species science, a new scientific paradigm, the beginnings of which were described by Charles Darwin more than 150 years ago. Investigations into the natural world no longer revolve around the question "How are humans different?" Instead, they cause us to wonder in awe at our relatedness. How we as humans think, feel, and behave is reflected not just in our mirrors, but in the faces of elephants like Happy.

2
A Delicate Network

Connie is a wild-caught female Asian elephant born around 1964. She has lived at the Dickerson Park Zoo in Springfield, Missouri, since 1981, when she was purchased from the Abilene Zoo in Texas. Since she was seventeen years old, Connie has undergone at least seventeen natural and artificial breeding attempts. She conceived three times as a result of mating with a bull, each experience with motherhood ending unhappily. One calf, born in June of 1985, was a stillborn female. Another female infant, Maiya, was born in 1991 and lived, but Connie rejected her. After six days, Connie took her back and all seemed well. But Maiya died at nineteen months from heart failure, the result of elephant herpes virus infection. In 2002 a third calf was born to Connie. She was named Asha. Mother and daughter were separated, though, when Asha was transferred to the Oklahoma City Zoo. Now Connie behaves aggressively to-

ward other female elephants, and she has developed a reputation for throwing feces, hay, and stones at onlookers. At these times, she is chained, and on at least on two occasions the restraints were kept on for several days.[1]

We left Happy and her compatriots at the Bronx Zoo, recipients of accolades and scientific acclaim for their role in helping trample down the idea that only humans possess self-awareness. Other animals have preceded the pachyderm in this status, but somehow the elephants' accomplishment brings an extra gravitas to the topic.

By explicitly extending recognition of species continuity to include the mind, we have the beginnings of a trans-species psychology, the study of mind and emotions that serves both humans and other animals. The elephant ethologists Cynthia Moss and Joyce Poole have been doing this all along, but mostly through "the outside in" of animal behavior rather than from psychology's "inside-out" perspective. Trans-species psychology allows us to imagine—without undue anxiety about anthropomorphism—what it might be like to walk in elephant "shoes" and experience what these awesome herbivores might be thinking and feeling, in much the same way that we think about ourselves and other people.

Does this mean elephants and people are the same? No, that would be simplistic. It makes no sense to say that any more than it does to say two people are the same. Refining a similar-but-not-the-same perspective requires delving further into neuroscience and psychology to better understand who we are, who elephants are, and what it is that confers simultaneously uniqueness and sameness. We begin from what is known to be held in common—the structures and mechanisms of brain and behavior—and then explore what is different, what are the things that make each person and elephant unique.

Scientists have long argued whether "nature" or "nurture" determines outcome. The nature school holds that inherited qualities predict psychological makeup, maintaining that much of behavior is innate, existing from birth and intrinsic in origin. Though knowledge is acquired, what is already stored in the brain and responsible for emotions, temperament, and actions is, according to this outlook, largely inherited.

In contrast, the nurture school maintains that behavior is determined not so much by inheritance but by who raises us, and how. Identities are shaped in the main by what we experience, not just by what we get from our ancestors. Adherents of the nurture school insist that each person is born largely a tabula rasa, a blank slate that becomes progressively etched by interactions with the surrounding world.

Opposing arguments dragged on over the years, favoring first one theory, then the other. Certain proof eluded both sides, and some data seemed to support each. On one hand, genes certainly emerge in similarities between relatives: familiar patterns of behavior and mannerisms ripple down successive generations, even when upbringing diverges. A woman who has never met her grandmother is surprised to find that they share the same habit of tapping the right foot when perplexed, and both are ardent eaters of peanut butter–and–onion sandwiches. But science has also shown that siblings raised in the same family may seem completely unrelated.

To help get to the bottom of the mystery, a series of studies was conducted to examine the habits and personalities of twins separated at birth and raised apart. Results showed that while in some cases, twins exhibit uncanny parallels, other pairs do not seem, except in physical appearance, to be any more related than passing strangers, and their personalities and temperaments could not be explained by either the family tree or their upbringing. Baffling, but as with many other conundrums, answers are neither black nor white, but a little of both biology and experience.

Research in the area of gene-by-environment interactions (GxE) reveals that inheritance and experience intertwine in ways that cannot be statistically separated, much less predicted. Neuroscience has also found that certain experiences can turn specific genes on or off. A child's experiences just after birth and even in the womb can influence the expression of her genetic potential. Each experience has an effect on what we think, feel, and act, and can have long-term effects ranging from the molecular to the behavioral.[2] The authors of the nearly six hundred–page tome *From Neurons to Neighborhoods,* commissioned by the Committee on Integrating the Science of Early Childhood Development of the National Research Council and Institute of Medicine, conclude that "The long-standing debate about the importance of nature *versus* nurture, considered as independent influences, is overly simplistic and scientifically obsolete. Scientists have shifted focus to take into account the fact that genetic and environmental influences work together in dynamic ways over the course of development." Furthermore, "Virtually every aspect of early human development, from the brain's evolving circuitry to the child's capacity for empathy, is affected by the environments and experiences that are encountered in a cumulative fashion, beginning early in the prenatal period and extending throughout the early childhood years."[3]

Psychologists have always considered the parent a critical influence for the child, but it was John Bowlby who brought attention to the critical role of the real, experienced interpersonal bond between parent and child. Bowlby was a psychologist who worked many years with homeless, orphaned, and abused children at the Tavistock clinic in England. In psychology's early days, students of the mind were preoccupied with exploring the dark waters of the unconscious and the ephemeral world of dreams. C. G. Jung, Sigmund Freud, and, later, other psychologists such as Melanie Klein, tended to explain children's emotional problems in terms of primordial drives of instincts and archetypes and their response to their environment. Bowlby brought a deeper, nuanced perspective to early relationships forged during infancy. Attachment processes—the visible and invisible transactions in the routine of baby care that other psychologists took for granted—took on greater dimensions in his theory. He demonstrated the powerful and lasting impacts of what happens between not just two objects but interactive psyches.[4]

Reminiscent of the mirror test in its ability to connect inner and outer, attachment theory provides an interpersonal map to track something as subjective and intangible as mother's love, and relate it to objective details of adult life and personality. The faces peering down as we lie in a crib and the arms that enfold us create an internal template that carries through adult life. As Bowlby himself wrote, "How a person construes the world about him and how he expects persons to whom he might become attached to behave . . . are derivatives of the representational models of his parents that he has built up during his childhood."[5] Childhood social experiences serve as a lens through which all manner of future relationships—friendships, romantic involvements, and work interactions—are perceived and guide how we act. We have also learned that relationship experiences are not limited to parent-child interactions. The psychologist Judith Harris showed that social influences extend well beyond the family and that peer group interactions can have equal, if not greater, effects.[6]

Bowlby's observations have since been confirmed by brain imagery: neuroimaging techniques such as fMRI (functional magnetic resonance imaging) make it possible to see *beneath* the "skull beneath the skin" and help to resolve the tangle of contradictions imposed by dichotomous Cartesian thinking. What happens on the outside—our relationships and interactions with the environment—are mirrored on the inside by neurobiological patterns and processes. Viewed through that scientific lens, the dreamy paintings of mother and child by the artists Mary Cassatt and Pierre-Auguste Renoir suggest the active sculpting of tiny neural pathways in the infant's brain that are stimulated by the cooing and cuddling of maternal embrace. Intimate gazes, smells, and sounds initiate a dialogue between the right hemispheres of caregiver's and child's brains, their emotional and social centers. This brain-to-brain conversation of loving touch and soothing sounds tunes the baby's emotional circuitry rather like the tuning of a musical instrument. By encouraging the brain's neurochemistry to percolate and flow, minute, imperceptible shifts in tone and color of maternal communications construct the very architecture of cells in the cerebral cortex. As Allan Schore, whose seminal work in neuropsychology has earned him the epithet of "American Bowlby," explains: "Nature's potential can be realized only as it is facilitated by nurture." Neither nature nor nurture alone is sufficient, but their synergy leaves us the same and different from each other.[7]

Current research paints a very different picture from Saint Augustine's portrait of the mind as an insular, uniquely human entity, separated from body and world like water held in an earthen vessel. Today's science has shown that the brain does not evolve in a void; instead, it is intrinsically relational, combining inheritance and acquisition to create a unique self. We are not, after all, the sole authors of our sense of who we are.

There is more. Not only is the brain social, but it is emotional. Historically, emotion was considered inferior to cognition, even extraneous to right thinking and living. Descartes and his followers celebrated the affectless mind and left passionate pursuits to artists and aesthetes. George Santayana provided perhaps one the most scathing assessments of emotions when he wrote: "Emotion is primarily about nothing, and much of it remains about nothing to the end."[8] From the perspective of science, though, emotional experiences are neither "about nothing" nor less important to day-to-day living than rational thought. Indeed, in the context of development and mental processing, they are formative ties that bind the mind. Far from being separate, emotions and cognition—like nurture and nature, mind and body—are partnered processes playing equally critical roles, even in territory conventionally considered best ruled by rationality. The neuroscientist Richard Davidson notes that "complex decisions—who to marry, which job to take—cannot be made solely on the basis of a cold calculus that involves the weighting of pros and cons in a formulaic prescription, rather, such decisions are typically made by consulting our 'feelings.'"[9] An entire field, affective neuroscience, now exists as a complement to cognitive neuroscience, and is devoted to understanding the neural substrates of emotions and feelings. This suggests that Descartes's famous *cogito ergo sum* (I think, therefore I am) needs updating to account for current scientific knowledge: "I feel (and think), therefore I am."

But are these recent disciplinary reunions pertinent to elephants? Do attachment and other psychobiological theories of self and emotional development apply to the likes of Happy, Echo, and the young South African bulls? The answer is a definite "yes," which further confirms the importance of Bowlby's work. By the early twentieth century, ethology and psychology were already diverging fields. Bowlby was one of the few to maintain a vital connection with the theories of the scholars of animal behavior—Robert Hinde, Niko Tinbergen, and Konrad Lorenz. Bowlby's keen observations of children, along with his knowledge of evolutionary biology, brought a rare fluidity to understanding behavior across species. Using diverse examples, Bowlby illustrated that bonding is present in all mammals and constitutes a phylogenetically evolved adaptive strategy found throughout the animal kingdom. In particular, he saw attachment as central to social obligates such as elephants and humans, whose lives revolve around and are mediated by family and friends.[10]

From birth onward, parrots, elephants, people, dolphins, chimpanzees, and myriad others depend on relationships to survive. Even for adults, life in the herd or flock is defined by the social web. Unlike chickens and other precocial species—those who start to eat and exhibit other survival behavior on their own just out of the egg or at birth—altricial birds are those whose young, as with mammals, cannot survive without a mother or other protector until they have matured. That is not to say that chickens are asocial, or that they lack the rich emotional lives characteristic of parrots. In fact, the traditional division between precocial and altricial classification is less a dichotomy than a continuum, and researchers have come to recognize that "behavioral development in all parts of the altricial-precocial spectrum is more flexible than originally thought."[11] Chickens and ducks, who practically pop out of the egg pecking, are as sensitive to their early surroundings and interactions as are elephants and macaws.

In basing psychological theory in ethology, Bowlby's work solidly grounded today's species-common unitary model of brain, behavior, and mind. The relational, emotional, and trans-species model, including mirror neurons, explain at least in part why humans have such an affinity with elephants and other animals—what E. O. Wilson calls "biophilia." Attachment theory also explains why elephants act like elephants and not completely like people, and conversely, why people grow up to be people, even when they share so much with pachyderms. We are who we are not just because we are born elephants or people but because we are schooled in distinctly different cultural ways. Hardly an earth-shattering revelation: an elephant is an elephant is an elephant. But the devil is in the details. Attachment theory, and its marriage with neuroscience, called interpersonal neurobiology, provides an operational approach to explaining species' differences and similarities. It also brings an appreciation for personality differences.

Elephants' personalities are as diverse as those of humans. In a pilot study investigating elephant personalities, the Amboseli Elephant Research Project (AERP) in Kenya found that a group of eleven female elephants "show consistent differences between individuals in standard measures of personality traits"; in fact, twenty-seven adjectives applied to describe the variety of individuals.[12] Dame Daphne Sheldrick has observed a rainbow of personalities and moods. Elephants, she writes, "can be happy or sad, volatile or placid. They display envy, jealousy, throw tantrums and are fiercely competitive, and they can develop hang-ups that are reflected in behaviour. . . . They grieve deeply for lost loved ones, even shedding tears and suffering depression. They have a sense of compassion that projects beyond their own kind and sometimes extends to others in distress. They help one another in adversity, miss an absent loved one, and when you know them really well, you can see that they even smile when having fun and [they] are happy."[13]

Dame Daphne is in a unique position to judge. She is a kind of latter-day elephant-world mix between John Bowlby and Mother Teresa. A native fourth-generation Kenyan born in 1934 and the founder of the David Sheldrick Wildlife Trust, named after her late husband, David Leslie Williams Sheldrick, Dame Daphne began taking in orphaned elephants at about the same time John Bowlby wrote his seminal works. The Trust is an intricate and complex institution. It comprises an elephant nursery located at the edge of Nairobi National Park and two rehabilitation centers in Tsavo National Park, where young elephants graduate from their nursery and begin their transition back into the wild.[14]

Dame Daphne's work started with the two-year-old orphans Samson and Fatuma. Samson lost his mother and family during a severe drought that decimated the robust Kenyan herds, whereas Fatuma's family fell victim to poaching. But it was not until 1987, shortly after the death of her husband, that Dame Daphne achieved what would become an internationally acclaimed success: rearing infant elephants. Since then her work continues to expand with the steady increase in young elephants in dire need of care. As of 2008 the Trust had successfully rescued and hand-reared eighty-two infant African elephant calves, two from the day of birth, and all younger than a year in age and fully milk dependent.[15] The Trust has provided an unrivaled up-front-and-personal view of infant elephant development and a window into the elephant soul.

Dame Daphne describes the baby Olmeg as "a complex character, deeply sensitive and easily wounded." On the other hand, Ajok is a "prankster with a sense of humour, a show-off and the most adventurous." Dika is "probably the 'nicest' character—very gentle, very sensitive, with an innate 'softness' yet with depths of hidden strength," while Edo is "rather shy and remote" and Lesanju is a "very loving, nurturing and responsible little elephant, old for her years."[16] Each one is special, with unique features and personality, yet all are distinctly elephant. Herein lies the clue to Dame Daphne's success.

Vital to infant elephant salvation has been her ability to raise them like elephants. This is borne not just in the lives saved but in the care she provides, which ensures that the traumatized orphans who are left in a void without others of their kind are brought back to their elephant roots and eventually reintegrated with wild society. All surviving orphans have been raised to become, or are on their way to becoming, full flourishing members of a culture as complex as it is vast.

Elephants are members of the order Proboscidea, so named in recognition of the remarkable and unique trunk that distinguishes them from all other animals. Historically, elephants roamed over immense distances. The rhino-killing youths were savannah African elephants *(Loxodonta africana)*, whose cousins are the forest elephant *(Loxodonta cyclotis)*; the Asian elephant *(Elephas maximus)*; and the Borneo pygmy elephant *(Elephas maximus borneensis)*.[17] Asian elephants evoke the same fascination as African elephants, but they hold dominion in the lands of India, Burma, Sri Lanka, Thailand, and other Asian nations.

The African elephant is the largest of the three, making the species the biggest land mammal. Adult males—bulls—may weigh up to 15,000 pounds, with a shoulder height averaging between ten and thirteen feet. Females tip the scales at 8,000 pounds and average some nine feet at the shoulder. Everything is big about elephants. Their tusks, which are actually incisors, can weigh more than 135 pounds and stretch to ten feet. Elephant ears can span five feet or more, their skin is two and a half inches thick, and they can live into their sixties. It takes almost two years' gestation before a 200- to 250-pound baby elephant comes into the world. Yet despite their formidable dimensions, elephants eschew aggression and are known for making friends with rhinoceroses rather than for killing them.[18] And something else, something that perhaps provides a more telling glimpse into their minds: elephants move with remarkable deftness and grace.

Elephant stealth is legendary in the stories hunters tell around the smoky brightness of the campfire: an intelligence and keenness of mind evokes an extra measure of uncertainty in the anxious human.[19] Baron Bror von Blixen-Finecke, husband of Karen Blixen, known more commonly as the author Isak Dinesen, wrote about elephants' tracking abilities and the quiet with which they move. A voracious professional big-game hunter after the First World War, the baron describes a fearful moment in the Congo when he became prey to the giant bull he sought: "Not a sound was heard—perhaps he had made off. Then I heard a faint crackling behind the tree against which I was leaning. I looked cautiously over my shoulder. Yes, there was the tip of his trunk swinging only a few yards from my feet."[20]

Dalene Matthee also writes about elephants. Her name is unfamiliar to most readers outside South Africa, but her books have been compared to the challenging realism of D. H. Lawrence and Thomas Hardy. Saul, the main character in *Circles in a Forest,* tells about his encounter with an elephant in the Knysna woods: "He heard nothing, not the snapping of a twig or even a movement. Nothing. He had just got off the footpath beyond the sled-path when he looked up and saw him. The elephant."[21]

Beyond their obvious intelligence, elephants are famous for their family life. Elephant society, like that of humans, is composed of an intricate, nested, and delicate network of relationships.[22] African elephants typically live in stable families comprising ten individuals on average. Sometimes families temporarily coalesce with others to form a larger group, a so-called fusion. Core social groups may persist for decades, and adult females remain with first-order relatives when any separation, or "fission," takes place.

While the majority of elephant research has focused on the females—the lure of baby elephants milling around giant wrinkled feet portrayed in so many films—Joyce Poole made one of the earliest forays into the impressive world of bull elephants to discover that they too had an intriguing subculture and language. Male elephant rearing is divided into two major phases.[23] From birth until the ages of nine to fifteen years, young male elephants live with their female siblings and cousins in the natal, family unit. Early life is rooted in female culture with female ways. But as they move from childhood to their teens, male calves gradually decrease physical closeness with their mothers, or are shooed out. At puberty, they leave their siblings and cousins and travel with other bachelors in a second phase of socialization, interrupted only by the occasional visit home.[24] Young bulls are initiated into the masculine world to learn the basics of elephant etiquette and the complexities of bull life in the great expanse throughout which they roam and must survive.

Bonding between and infant and his mother occurs early on, and a young elephant may suckle from his mother until about four years of age. Yet while it is the mother with whom the infant interacts the most, a young elephant has more exchanges cumulatively with other family members: elephant life is communal.[25] The elephant care village is made up of "allomothers"—other females, siblings, aunts, and cousins who tend to every aspect of calf upbringing, though on the rare occasion when a baby tries to suckle from another calf's mother, that female usually rejects the trespasser. Social exchanges include such physical greetings as touching trunks on different parts of the body and mouth or rubbing against each other. Touching also provides comfort and assurance when a baby becomes distressed from straying too far afield.[26]

Elephant society includes a strong sense of belonging and family identification. This awareness, sense of connection, placement, and relatedness extend beyond the family unit. One expression of this unity is a strong value of reciprocity, and to a certain extent a kind of leveling that marks relationships. For example, the infant who has been tended by young females, will, when their allomothers bear their own babies, reciprocate by caring for these new young family members. Regardless of genetic relatedness, rank, or gender, all elephants seem to show concern about ill or dying individuals. The literature is replete with examples. Elephants have been observed gathering around and trying to rouse an ailing elephant to his or her feet, as we saw with Ely, who was born with poorly articulated carpal joints.

The matrix of herd life is also a place where ecological and social knowledge is gained. Learning and knowledge extend beyond the inner dynamics of the family. On average, a single family unit will annually encounter 25 other families (or roughly 175 adult females), so they must be able to recognize and communicate with each. Indeed, playback experiments have demonstrated that an adult female may be familiar with the contact calls of around 100 others in the population.[27]

Knowledge relating to geography and events—where and when specific foods can be had, for instance—are reflective of a relational geometry, in which the matriarch represents the apex of knowledge. For the young, objective knowledge is gathered by direct trial-and-error experience mediated through older, more knowledgeable family members. For example, at about three months of age, young elephants, still nursing, start to browse vegetation. The student will reach up with her trunk to the mouth of a family member to touch and taste what her older relation is eating; thus she learns what is edible and how it tastes and smells. Learning how to drink and manipulate (trunkulate?) the trunk is also a matter of observation and practice. Play forms a central activity of young, growing elephants and, beyond sheer joy, serves to establish relatedness and autonomy.

Compared with most other species, elephants are extremely well studied, and a tremendous volume of data has been collected from field observations.[28] As a result, many aspects of elephant society have been made transparent. However, subjective knowledge is also a necessary ingredient in the formula for successful human rearing of an elephant orphan. All of Bowlby's nuanced infant-parent communications must be emulated, adapted to elephant culture, and planted into the Sheldrick orphans' heads if they are to survive outside the Trust compound. Without Dame Daphne's thoughtful care and extraordinary knowledge of elephant culture, the majority of rescued infants probably would have perished either in the progress of their rehabilitation or after their reintroduction into the bush with their wild compatriots. Indeed, despite the sterling skills of Dame Daphne and her Elephant Keepers, some infant elephants arrive already beyond even her capabilities and tragically succumb.

The evolving mind absorbs like a cognitive and emotional sponge. During the period when a child is heavily or even completely dependent on her family to survive, the brain is considered to be at its most "plastic"—most receptive to the environment. Such receptivity permits successive generations to keep up with rapid environmental changes, but it can also create vulnerability if what a child learns meshes inadequately with the adult environment. It often happens in wildlife rehabilitation that an animal raised by humans develops a "bicultural" identity. Such a cross-fostered individual (one reared by a member of a different species) may acquire too much human culture or too little of her species of origin to make a successful transition from human midwifed care to independent living with conspecifics. The outcome can be painful, even fatal.[29] A reintroduced cub or fledgling may not have learned the requisite cultural skills of a youth taught by conspecific elders. Some newcomers commit unwitting trespass because they have not learned social etiquette or have been released near a group other than their own kin: the group may reject a youth for violating rules and territory. A similar problem is encountered with individuals such as Billy Jo, a male chimpanzee who lived in sanctuary at the Fauna Foundation outside Montreal.[30]

Chimpanzees and other nonhuman primates in captivity are either captured from the wild or captive-bred. In either case, the individual is usually reared by humans or someone other than his or her birth mother and family. Primate captivity and breeding serve to provide exotic pets, human-surrogate experimental subjects for laboratories, and performers for entertainers.[31] It is not known whether Billy Jo was wild-caught or born in captivity. Clearly, though, he was forcibly taken from his mother and raised by humans and allowed little if any interaction with other chimpanzees.

Until he was a teenager, Billy grew up much like any young boy might: going fishing, hanging out, eating ice cream and junk foods when not caged.[32] Suddenly, at the age of fifteen, he was sold to a biomedical laboratory, where he served the next decade and half as an experimental subject undergoing more than two hundred anesthesia "knockouts," followed by painful medical procedures. He lived alone in a suspended cage lined up with other chimpanzees similarly confined, and instead of his former regimen of intimacy with humans, he encountered only laboratory personnel covered in white protective garb, including masks, bearing anesthesia dart guns or doling out food through the cage bars.

When Billy finally was released into sanctuary, his conditions improved dramatically. However, he was unable to mobilize one important resource in psychological recovery from trauma: social support. Although potential chimpanzee friends were available in the sanctuary, Billy did not know how to appropriately socialize with other chimpanzees, and they, for their part, could not understand him. Because of his upbringing, Billy was attuned to human habits and culture, not those of chimpanzees. He eschewed foods favored by chimpanzees—leaves, branches, hard-shelled nuts—in favor of popcorn, pizza, and other snack foods and beverages. These delights were far more interesting than the nutritious foods offered to him. Unlike other chimpanzees who usually wait and inspect this less-than-typical fare, Billy would immediately bite into these goodies without hesitation.

He used a plate and plastic utensils instead of eating with his hands, which most other chimpanzees do, and enjoyed "twirling" spaghetti on his plastic fork. Billy completed each meal by taking a napkin and carefully wiping his mouth and chin. When given the option, he chose to pour all his drinks into a paper cup and save it for future use. If he saw someone with a cup of coffee, he would ask for some for himself—gesturing for cream, sugar packets, stir stick, and napkin—then pour and stir the ingredients into the coffee, gently place the stirrer down, and take a slow sip. He showed other human proclivities.

Billy was obsessed with washing his hands and face, and after pointing to a box of tissues, he would carefully lift and use one to blow his nose. He enjoyed dressing in human clothes such as baseball caps and shirts, taking care to put them on straight. He also enjoyed flipping through magazines, especially those with photos of human women, and enjoyed painting with different colors. However, not all of Billy's human-derived habits and ways brought pleasure, and the uneasy identity born of cross-fostering created profound difficulties.

Encounters with other chimpanzees, even though carefully orchestrated by the sanctuary director, continued to fail, and eventually Billy was severely attacked by the others. After the incident, he would lock his door at night and check it several times to confirm that it was secured. Not only did he suffer from the loss of his early family and social group, but neurobiologically, psychologically, and emotionally, he was attuned to human ways. Chimpanzee ways were foreign. This life in limbo—perhaps what might be called chimpanzee in body and human in mind, or dual identity—caused intense periods of depression and sadness that the sanctuary was able to address only in part. Cared for and loved by human sanctuary caregivers, Billy nonetheless remained suspended between cultures.

In other cases, intervention by another species can be the saving grace. Take Owen, a young hippopotamus orphaned by the tsunami in Kenya, who was taken in by Mzee, an ancient tortoise: a picture of an odd couple if ever there was one. International news carried endearing pictures of the young hippopotamus gazing lovingly up into the weathered-looking face of the tortoise, a partnership that ended only when humans took away Owen to be with others of his species.

Then there is the injured crow whom the ethologist Konrad Lorenz nursed back to health. When the crow recovered and matured, he fell in love with a Swiss woman from a nearby village, ignoring all bird companions. Eventually, his love spurned, the crow returned to Lorenz's home, still indifferent to his fellow corvids. Recently, a couple filmed another love-struck crow from their neighborhood who saved and cared for a young stray cat. Diligently and tenderly, the crow brought food, even worms, to nourish the fragile feline. Over time, the two began to play together, enjoying a newfound companionship.

Across the ocean, Jessica the hippopotamus, a member of a species considered to be one of the most dangerous, was rescued by the Jouberts, who found her alone on the banks of the Blyde River in South Africa. Jessica probably had become separated from her mother during devastating floods and had washed downstream. Through meticulous nursing by her human rescuers, Jessica recovered from her trauma to become part of the family. A visit to the Jouberts' often finds Jessica, after a hippo-style swim, ambling her bulky frame into the kitchen, where she is hand-fed fruit; later she is tucked into bed on the porch next to Za Za the Rottweiler. The Jouberts speak frankly about their relationship with her: "We realized that we were now her parents." Film clips of this trans-species family attest to their mutual commitment.[33]

Interspecies love goes both ways. Feral children, orphaned or abandoned, have been taken in by families of other species. Rudyard Kipling's *Jungle Book* tells of the boy Mowgli, who was nursed and mentored by wolves, snakes, and leopards, and François Truffaut's film *L'enfant sauvage,* describes the "wild child" found roaming the woods of southern France. Some children become attached to the wolves or dogs with whom they have lived to the point of choosing their adoptive species over humans when they are forced to return to "civilization."[34]

Neuroscience shows why, while looking so different on the outside, elephants and people are much alike underneath, and attachment theory shows how what we inherit interacts with what we experience. Internal and external processes of bonding may be common across species, but variations in how and by whom we are raised are responsible for the many variations on the theme of being an animal. This knowledge brings us one step closer to understanding why at times elephants may not act very pachyderm, as in the case of the rhino-killing bulls, and why their naturally gentle behavior has come to resemble aspects of humanity that we would rather not own.

3

A Strange Kind of Animal

Dawn comes slowly on the veldt. With the touch of morning light, each still-life character wakens into slow motion. But there is nothing measured in the young bull elephant racing toward the feeding rhinoceros—a dull gray mound grazing among gilded yellow grass. Hornbill chatter and the gazelles' tentative glide burst into loud squawks and frantic movement as the two giant mammals collide. The air explodes with the crash of body against body, gray against gray, and deafening bellows. In a short while, stillness descends. Gradually, birds and antelope filter back, and the landscape resumes its former repose. One by one, vultures drop from the sky to begin methodically picking over the armored rhino corpse before hyenas take over. It is South Africa, 1992, two years before prisoner Mandela would become President Mandela. Within a decade,

more than one hundred rhinoceroses would be dead, and the bull elephants gunned down as culprits.

Elsewhere, dawn reveals another grisly scene. Hornbill chatter is replaced by commuter traffic, and instead of antelopes, lawnmowers skate over spacious green lawns of a Florida retirement community. The day's heat is starting to be felt, and the weight of humid air has begun to press. Nothing obvious suggests that the van outside the suburban home is anything extraordinary, yet when the doors slide open, instead of the usual golf-clad retiree, three young men jump out and disappear inside the house. In a few minutes, the neighbors hear loud gunshots, and the sound of the van speeding away. Police find sixty-eight-year-old James Miller and his sixty-three-year-old wife on the floor. Mrs. Diana Miller is dead from two gunshot wounds to her chest and head. Mr. Miller is still alive, despite a bullet that has grazed his head and penetrated his ear. The prosecutor asks for the death penalty in two cases and consecutive life sentences with no parole for the youngest, the alleged shooter.[1]

What possible connection could there be between a murderous elephant and a murderous human separated by thousands of miles? Humans may kill with disturbing frequency, but historically, inter- or intraspecies violence is uncommon among elephants. Even during musth, the period of heightened sexuality, when male elephants become openly aggressive as elevated testosterone charges through their systems and their faces become marked by the distinctive stream of temporal gland secretion, male-on-male injury and mortality are relatively rare. The string of more than one hundred rhinoceros fatalities took on greater significance in light of news from other parks and reserves.[2] There were similar incidents involving rhinoceroses at other locales, and at Addo Elephant National Park, also in South Africa, male-on-male elephant aggression is responsible for 70 to 90 percent of adult male deaths.[3] These statistics stand out as exceptions to the natural history rule of elephants.

Keith Lindsay, a researcher in the Amboseli Elephant Research Project, notes:

> The detailed studies over more than 30 continuous years have recorded only four cases of males killing other males during musth-related contests over females. The great majority of male-male contests result in one or the other male backing down, after recognizing the superior strength or motivation of his opponent; the escalated contests that result in mortality occur when the two males are most evenly matched. This sophisticated ability to assess the likelihood of winning a fight or sustaining serious injury has been developed over many years of experience in male society, and dysfunctional elephants who rapidly escalate to violence are clearly lacking this social learning.[4]

Today, violent outbursts in elephants are not anomalous.[5] Elephant clashes with other species are not limited to rhinoceroses. The extent of discord between elephant and people has warranted the coining of the acronym HEC, for "human-elephant conflict." In Nepal killings of humans by herbivorous elephants outnumber those by predator tigers.

Elsewhere in Africa, such as in Sierra Leone, where humans and elephants have a history of peaceful coexistence, some three hundred villagers left their homes because of "unprovoked" elephant attacks and damage. Disturbingly, new patterns of aggression are not limited to the male of the species. Female elephants have been observed to uncharacteristically leave their calves vulnerable in order to charge tourists.[6] The ethologists Delia and Mark Owens, authors of *Cry of the Kalahari,* have noted that some mother elephants ignore their distressed or endangered infants. The cries of young who have strayed from the herd's protection go unheeded. Other behavioral oddities include social aloofness among herd members, something else completely uncharacteristic of the close-knit family-values-oriented elephant society, in which group members retain their deep connection with frequent affectionate caresses and thoughtful inquiries by serpentine trunks.[7]

According to the forest conservator in the Wildlife Circle of Doranda in India, increased human populations and reduced elephant habitat have caused the dramatic rise in human-elephant conflict: "As forests become more fragmented and degraded or are converted to monoculture plantations, both elephant feeding and migratory patterns are disrupted. The results are sadly predictable. A herd of 13–18 elephants killed seven people during one five month period. Another herd of about 60 elephants killed 11 people in 1988 and another 12 in 1989."[8] The inspector general of forests and director of Project Elephant in India echoes the same concern. There were sixty-seven human deaths in 2006–7 and ninety-one deaths in 2005–6 in the Jharkhand region as a result of "conflict with wild-elephants." These numbers come to an annual average of 250 deaths when all eastern states are totaled. Elephant fatalities are even higher: "nearly 700 elephants died in Orissa between 1990 and 2008. . . . 34 per cent of the elephants died due to poaching, 24 per cent died as a result of accidents including electrocution and road or railway mishaps."[9]

Statistics and testimonies can no longer be argued away as unfortunate, aberrational events or whimsical stories. People who have lived with elephants for generations report "rampant conflict," "serious trouble with wild elephant herds," and "angry wild elephants . . . attacking the locals."[10] Elephant behavior has changed. A forest worker, Debojeet Saikia, says, "Earlier if they saw us foresters, they would run. But now they don't care even if we open fire or burst bombs. We could scare them off with search lights but now they come after us."[11]

On June 30, 2001, more than sixty people from twelve villages met in the Meshenani area of the Olgului/Ololarrashie ranch, "the largest, most important communal land that almost engulfs the Amboseli National Park." Participants described "the peaceful coexistence of Maasai and wildlife in the delicate balance of the ecosystems within which they live." A Maasai elder in Kenya maintains that in the past, "elephants hardly ever attacked people unless provoked, thirsty or instinctively reacting to an experience of past attack," but as a result of threats coming from "commercial agricultural expansion; sidelining of the Maasai from mainstream nature conservation; insensitive tourism practices; and continued loss of Maasai traditional lands to other modern economic enterprises," elephant-human relationships have changed. The Maasai insisted that the "ongoing destruction of forests, commercial hunting, and loss of wildlife migratory routes and breeding grounds must be stopped now if the future of wildlife in Kenya and Tanzania is to be guaranteed." Without such steps, "we lose land and culture [and] elephants and other wildlife lose habitat."[12]

In light of reports from Asia and Africa, the South African bulls turn out to be the tip of a very unsettling iceberg, or, to apply another metaphor, they represent a several-ton canary in the environmental coalmine. By accumulation of data over a broad geographical expanse, isolated events begin to look like evidence of a behavioral sea change in normative elephant behavior. Since attachment theory and neuroscience have been helpful in understanding what makes an elephant an elephant in mind and body, it makes sense to revisit these subjects and see what causes the reverse: what makes an elephant act unlike an elephant, uncharacteristically "violent"?

Violence is a powerful word, and it is not usually employed in the case of animals, let alone herbivore elephants. In contrast to the more frequent animal descriptor of aggression, violence includes intent and implies moral violation, attributes typically reserved for the human species. As one well-known pair of forensic sociologists put it, "No one imagines that the plant has intentionality."[13] Conventional thinking would declare rhododendrons and elephants incapable of violence because neither is supposed to have the requisite higher-order functions associated with a sense of morality or the mental sophistication that permits action other than instinct. Animals are expected to kill, and the drive to do so is a compulsion of base instinct. Killing is what lions, tigers, and bears do for a living, almost mechanically, without emotion.

Perceptions of animals and emotions are linked. Like species, emotions have been ranked according to evolutionary progress, *scale natura,* or the Great Chain of Being. Time separates humans from animals as much as cognition and emotions do. In the past brain structure and functions were thought to be layered upon each other like geologic strata where "The various lines of psychic development start from one common stock whose roots reach back into the most distant past."[14] Accordingly, "it should be possible to peel the collective unconscious, layer by layer, until we come to the psychology of the worm, and of even the amoeba."[15] Humans may share the instincts of worms and snakes, but it is the human veneer of higher faculties that prevents William James's unthinking blind ferocity from taking over. More than a century later, the psychiatrist James Gilligan, who has studied human violence extensively, voices a similar sentiment: "It is not a coincidence that our human propensity to create morality and civilization, and to commit homicide and suicide, are the two characteristics that most specifically differentiate us from all other species."[16]

Violent emotions, the instinctive, involuntary reactions, only "upset the rational order of consciousness by their elemental outbursts," breaking through this veneer to reveal a common animal ancestry.[17] Violent anger is an amoral emotion because it can be clearly seen in rats, dogs, toddlers, and other creatures considered to lack a well-developed moral life.[18] Aggression, anger, and rage are shunned in humans, while their antitheses—humanity, love, compassion, and rational thought—are celebrated, not only because they are obviously more desirable but because their opposites are considered to be a regression to animal baseness: a manifestation of some human pathology or failure to repress a destructive impulse. (Ironically, humans alone have been allowed the privilege of mental disorder.)

Yet when cloaked in the collective righteousness of human war and other sanctioned injurious practices, violence becomes more acceptable. In these instances, violence loses its purplish cast and settles into muted grays of rationality. Anglo-American common law considers that killing in response to discovery of a wife's infidelity is an act of a "reasonable man": it is not condoned, but it merits a reduced penalty. Violent sexual jealousy is deemed normal or at least unsurprising both in societies in which the cuckold's violence is seen as a reprehensible loss of control and in those where it is seen as a praiseworthy redemption of honor.[19] Blame is mitigated by sympathy.

Animal and human become strangely blurred, then, with the juxtaposition of *violence* and *elephant,* and with the phrase "human-elephant conflict," which situates the species on equal, albeit combative, footing. One of the first times that scientific models of human and nonhuman animal behavior crossed, and animals were ascribed humanlike attributes, occurred when "war" between two groups of chimpanzees in Gombe National Park, Tanzania, was reported by Jane Goodall.[20] Despite a desire by many other scientists to discredit her on the grounds of anthropomorphic projection, species parallels were marked. Chimpanzees were observed engaging in multiple, systematic gang- or group-led killings, infanticide, and finally the eradication of one clan by the other (what one might call, in the case of humans, ethnic cleansing), finally leading to peaceful coexistence.

Since then, science has documented additional examples of violent behavior, including—contrary to Gilligan's assertion that the behavior is uniquely human—suicide. Intentional self-harm leading to death—inward-directed violence—has been observed in animals, but has rarely been called suicide.[21] Nonetheless, it is not unusual for some species in captivity to die from self-inflicted injuries. Moluccan cockatoos in captivity, after losing a human caregiver to whom they have bonded, sometimes pick at their chests until they expose the bone and then succumb from these wounds, and as we shall see, Jenny, an elephant at the Dallas Zoo, has been seriously damaged from repeated self-injurious behavior.

Not everyone has considered the lines between animal and human affective behavior so distinctly drawn. David Morris muses on modern culture's ambivalence in this regard. "We are a strange kind of animal. We have bodies like other animal bodies, and move like them. Our stories tell of these human-animal affinities, we speak of animals as totems or familiars, and we elucidate principles of the cosmos or society in terms of animal behaviour. Yet, in telling, speaking and elucidating—indeed in building and thinking—we find ourselves estranged from the animals."[22]

Even the ethologists Konrad Lorenz and Niko Tinbergen have openly criticized scientific models for being prejudiced. Lorenz argued that aggression in the animal kingdom is normative, but human perceptions made it perverse. What may seem to the human observer disturbing behavior in an animal is in actuality "misguided by sensationalism in press and film, [whose authors imagine] the relationship between the various 'wild beasts of the jungle' to be a bloodthirsty struggle, all against all, [but] such things never occur under natural conditions. What advantage would one of those animals gain from exterminating the other?" In Lorenz's eyes, killing is part of life, "appearing either appropriate and self-protective, even constructive, as in healthy self-assertiveness," whereas in humans aggression assumes its pathological aspects, becoming violence when it is "inappropriate and destructive."[23]

Compared with natural disasters, to say nothing of human-against-human violence, deaths caused by elephants are negligible. Amid the chaos gripping the world today, one might expect that elephant aberrance would get lost, a tiny signal against an obliterating white noise of human distress. But elephant incidents receive far more attention than they warrant statistically. There appears almost a kind of eagerness to prove that elephants are "as bad as us," while simultaneously denying that likeness demands comparable rights.

Labels other than *violence* or even *conflict* are available to describe elephant aberrance, and the preference for these charged terms is informative. *Violence* serves the purpose of media out to grab attention, but there may be a more psychologically significant reason for using the word. Intentionality implies directionality. Directionality implies motion, suggesting that elephants are stepping out of their assigned role as part of the passive landscape against which human dramas play out. In this new scenario, elephants have become actors, and in so doing, have upset humans' sense of order. Elephants have trespassed against physical and psychological boundaries and have flouted human privilege by damaging houses, consuming crops, and showing a commitment to kill: behaviors legitimate only in human circles.

Suddenly daily reports from Assam, Kerala, Tanzania, and Kenya are laced with a sense of outrage. Elephant violence invites human retaliation perhaps because elephants' apparent appropriation of human privilege signals a refusal to obey the rules of fair play, and therefore a type of guilt.[24] An elephant who kills a human is not only morally dubious but morally stigmatized.[25] Before, an elephant's charge was explained as a fearsome display by a mighty rogue bull. Now it is called "elephant rage," and bulls are brought down in India much like renegades.[26]

Neighbor turns to foe, salutations twist into hostile bellows and deadly cudgels. All of a sudden, humans have paired themselves with elephants in hostile bondage: the Arab-Israeli conflict, the War Between the States, Indian-Pakistani tensions, the human-elephant conflict. However, it is an unfair pairing; something has gotten lost in translation. The discovery of similarities should not obscure profound differences. Elephant culture and values are not the same as modern human culture and values. Elephants are not armed with deadly technologies. It is easy to forget that their "violence" exists only because humans have manufactured an environment that leaves them few behavioral choices, including those that seem to resemble human behaviors.

Below the surface, in the realm of neurons, there is yet another indication of how close humans and elephants really are. Biology suggests that either the young rhino-killing elephants don't come from elephants, or they aren't being raised like elephants. Barring any *Star Trek* visitations from other planets, attachment theory predicts that differences in elephant behavior are organic and native to the circumstances in which they were raised, circumstances that do not conform to those of the past.

Robert Hinde, an ethologist and colleague of John Bowlby, wrote that behavior "can be understood only in terms of a continuing dialectic between an active and changing organism and an active and changing environment with cause and consequence closely interwoven."[27] If we understand the concept of dialogue figuratively, Hinde's picture of behavior directly describes the ways the lives of elephants and people of Africa have changed: the two species are engaged in a very different sort of conversation since colonial occupation.

Everyday life of elephants in Africa and Asia no longer entails foraging and family life peppered by the occasional threat of a marauding predator. Now these preoccupations are dominated by a new factor: a magnified threat from humans that heightens everyday perils. Today's humans do not offer the terror of a single spear: they descend in helicopters wielding machine guns, bringing horrors of apocalyptic proportions. The skies—source of the gift of much needed rain in arid lands, and the lacework of birds carrying news of faraway plains and mountains—are no longer benign. Modern humans have the power to replace an entire wooded valley with crops or to change the flow and course of water that quenched the thirsts and cooled the bodies of elephants for generations.

As the land and ways of the African savannah and Asian jungles have changed, so have the conditions under which baby elephant brains develop. The rupture of elephant lands, lives, and history is mirrored in the legacy of rupture in their brains, bodies, and behavior. The object of our inquiry is therefore not just the elephants, not just the environment, but both, the dialogical space in between, which turns out to be the site of trauma.

Trauma is a specific kind of stress. There are all types of stress. We get stressed when we get stuck in rush-hour traffic. We get stressed when a lover announces she is leaving. We get stressed when the boss walks in the office and says that the deadline is moved up three hours for a job that takes six, and the list goes on. But while the hustle and bustle and uncertainty of the twenty-first century may be blamed for many ills, stress by itself it is not always bad. Stress is part of everyday life and survival.

In its broadest definition, stress can be thought of as what results from the difference between what we expect in body and mind and what we actually experience. Stress becomes harmful only when our expectations are not met by the environment in which we live and we are unable to adjust to the actual conditions. Every species (and individual) has its own psychological, emotional, and physical "envelope of tolerance" to which each has adapted and functions well. To a greater or lesser extent, everyone is vulnerable to the same stressors—the fear of death and threats, whether in the form of a lion or of an angry supervisor. But just as each person and animal has a unique body type, each has a unique psychological type. An elephant's body and mind are equipped to handle environmental stress differently than are the body and mind of a lion or a turtle or human. A penguin doesn't feel stress in subfreezing weather as a lion would, and a lion doesn't feel stress in the oppressive heat of the savannah as a penguin would. As the neuroscientist Robert Sapolsky notes, this is the reason why zebras don't get ulcers and we do: ulcers and other ailments related to a "strain in the nervous system"—that is, things related to severe emotional stress—are widespread in humans because of the postmodern life of traffic jams, multitasking, multimedia, and Super-momming. "When left to their own devices, zebras don't get ulcers." But when they undergo "severe and unnatural stress (e.g., when they are first transported into a zoo)," ulcers can and do develop.[28] If a psychologically or physically intolerable situation cannot be avoided, fled, or resolved in some way, if "we are no longer able to change a situation," stress emerges and "we are challenged to change ourselves."[29]

The inability to compensate for radical changes in the environment affects us at multiple levels and in multiple ways. Severe stress causes an overactivation of the hypothalamic-pituitary-adrenal (HPA) axis. This set of structures located in the brain functions as the body's neuroendocrinal chassis for dealing with environmental change and regulating stress. When something startles us—a sudden, loud noise or someone coming at us with a knife—we feel a spurt of adrenalin. Blood pumps, heart races, the flight, fight, or freeze program kicks in. It is the HPA axis that functions as the action-reaction boiler room, taking information in for mind and body and processing it to appropriately match what is needed to survive. The system is efficient and effective—unless the threat does not go away or is too big for us to handle.

Chronic stress, when the pressure does not let up, and trauma, when forces of danger are overwhelming, overexcite this regulatory engine such that it stays in high gear much like a Volkswagen Beetle trying to pass a semi on the freeway. First it starts accelerating in anticipation of the Big Truck; gets into the left lane and keeps accelerating to overtake the Big Truck; then, after gliding back into the right lane, safely ahead, slows down to a nice, leisurely VW pace—but in the chronic stress scenario, it can't. The gears are stuck and it stays in fourth gear even when it gets off the freeway to slow down.

After a while, running in high becomes taxing and wearing, a worrisome state for the nervous mechanic on the side. When this happens in the body, getting into high gear means circulating the various biochemical and hormonal packets, endogenous corticosteroids. Constant stress keeps them getting released and running throughout the body, all of which cause a change in biochemistry that eventually translates to what we see on the outside—a change in behavior and personality. We get grumpy, irritated, edgy, fearful, anxious, and fall prey to a myriad of other unhappy moods, illness, even memory loss.

The envelope of coping tolerance also relates to ontogeny—the physical, emotional, and psychological conditions that an individual experiences as he develops. As we learned from attachment theory, parents and caregivers form a matrix that communicates and interprets meaning and significance of the World Out There to the youngster. This is how an infant begins to learn about the world in which she will eventually live, mature, and have her own children. For better or worse, early relational dances influence perception, knowledge, and behavior.[30] The Nobel Prize winner in literature François Mauriac put it most succinctly: "We are moulded and remoulded by those who have loved us; and though the love may pass, we are nevertheless their work, for good or ill."[31] Consequently, what happens to the mother and family happens to the child's world and mind. When a mother is killed or the child is harmed, the world causes pain. The young body and mind adapt and shape to these environmental conditions accordingly. Thus what goes around, stressful interactions with Mum and Dad, comes around: stressful childhood leads to stressed behavior in maturity, and passes to the next generation.[32]

Traumatic disruption from a single event can create lifelong changes in personality and neural organization; in some cases, stress can even be transmitted from mother to fetus within the womb and permeate as deeply as genes. Hormones can turn genes on and off, so in a sense, the genetics of one twin, who had a loving, happy-go-lucky childhood, may function somewhat differently from that of his sibling, who grew up in a violent and unnurturing family. Abuse, neglect, or sudden parental death can impair the expression of those genes involved in neurogenesis and synaptogenesis of highly receptive developing brain circuits that inform memory, cognition, and emotions. What and how we remember, think, and feel is affected by traumatic experience. The impact of an emotional event such as the violent loss of a parent "affectively burns" into the primordial emotional and social centers of the limbic system, HPA axis, and the right frontal lobe of the brain—in animals and humans alike.[33]

What we now know from neuroscience was documented well before in study after study on behavioral effects of altered socialization and rearing of animals. Dr. Harry Harlow (with Stephen Suomi, whom he mentored and who is now chief of the Laboratory of Comparative Ethology at the National Institute of Child Health and Human Development) even devised apparatuses specifically designed to examine how changes in natural infant-rearing patterns affected both mother and child.

Eschewing the euphemisms typical of much contemporary dialogue, Harlow frankly described his work and his aims: "The only thing I care about is whether a monkey will turn out a property I can publish. I don't have any love for them. I never have. I don't really like animals. I despise cats. I hate dogs."[34] He designed a "rape rack" to force monkeys to mate, a "well of despair" in which infant monkeys were kept isolated in the dark from birth to one year. Using an "iron maiden"—a metal mother-monkey-shaped doll that stabbed or blasted cold air with sufficient force to knock a baby down—Harlow tested an infant's loyalty to his mother.

Today Harlow's studies and attitudes are decried, but his successors pursue similar research, taking care to describe their experiments in more politic language: "terminate" instead of "kill," for example.[35]

Harlow's work, as distasteful as it may be to most mainstream researchers, serves as a sort of grotesque funhouse mirror of current science. In fact, much of the research about attachment was implicitly framed from the perspective of the sort of trauma that is explicit in the design of his studies of relationships and the behavior they induce. In learning about love through its horrific violation, Harlow established that the importance of relationships goes beyond their role as conduits of food and other commodities.

Bowlby's lessons about early bonding—learned without excessive and brutal methods—also derived from the perspective of rupture, what he called loss and separation. Indeed, attachment theory might very well have been dubbed trauma theory. Many of the concepts underlying attachment theory came from observations of children suffering from traumatic compromise. One book in Bowlby's trilogy on child development is devoted to the effect of parent-child separation. Bowlby's work was motivated by his compassion for the plight of children left bereft of parental love. He writes that that "no one can be unmoved" by their stories and conditions and that the "loss of mother figure, either by itself or in combination of other variables yet to be clearly identified, is capable of generating responses and processes that are of the greatest interest to psychopathology."[36]

Mindful of the psychic costs that he experienced when sent off to boarding school at a vulnerable age, Bowlby regarded the fear and anxiety associated with traumatic separation as defining events. Separation events "occur so commonly in the lives of children, adolescents, and adults, and constitute so large a proportion of the major stressors about which we know, that a clear understanding of their effects is of immediate help to clinicians whose task it is to understand psychiatric disability, to treat it, and whenever possible, to prevent it." If the process of loving bonding is absent or interrupted, Bowlby continues, then the infant's learning path is altered. Loss, abuse, and abandonment are responsible for the profound corrosion of security and well-being, and for the resultant behaviors. Disruptions to early bonding cause acute stress, whose effects endure into adulthood. "One reason for this belief was that the responses and processes observed seemed to be the same as those found to be active in older individuals who are still disturbed by separations they have suffered in early life."[37] Indeed, Bowlby maintains that loss of the attachment figure is the principal agent in the development of psychopathology.

We now see that attachment theory is directly associated with traumatology. Attachment and trauma are two sides of the same theoretical coin, and the two together provide a new conceptual framework for understanding human experience and psychological life. Attachment studies how relationships, particularly those of infants and children, influence thinking, feeling, and behavior; traumatology studies how stressful disruptions and experiences, particularly those effected through our interactions with others, influence thinking, feeling, and behavior. Ruptures to primary bonds constitute relational trauma (in contrast to shock trauma, which might occur during war when witnessing an atrocity), which in symmetry with attachment processes, can be related and mapped through current models of the brain to their expression in behavior and psychological states. Trauma challenges or undoes what attachment creates.

However, while joined in theory at the levels of neurons and genes, the two frameworks hold very different places in science and society; the reason is tied to politics and perception. Harlow and company aside, attachment attracts with its heartwarming images of mother love and babies. On the other hand, trauma repels with images of death, violence, suffering, and controversy. The trauma psychiatrist Judith Herman writes in her seminal book on trauma, *Trauma and Recovery,* of the unique role and peculiar history of trauma studies. It is a history of "episodic amnesia": the field advances in fits and starts not because the subject is vulnerable to academic mood swings or fads but because it "provokes such intense controversy that it periodically becomes anathema."[38]

Part of the controversy derives from an intrinsic unease with the topic of suffering. Cognitive acrobatics may entrance, but mental distress attracts uncomfortable attention, sometimes even inspiring an impulse to conceal findings from a squeamish public, things that, like sleeping dogs, are best left lying.

Understanding the history of human traumatology and its place in science and society is important to our exploration of aberrant elephant behavior because of its intrinsically relational and therefore unavoidably political nature. An external, causal agent has created the trauma. Diagnosis and treatment implicitly involve more than the victim; they involve cause: the person or circumstances that have inflicted physical or psychological injury. This relational framing of psychological suffering departs from conventional models because what and how a person thinks, feels, and acts is acknowledged to be vulnerable to other events or people.

In childhood, the agent of trauma may be something or someone who disrupts what Bowlby refers to as the "warm loving" bond between parent and infant. But the relational aspects of trauma and its impacts extend well beyond childhood to the broader web of social relationships and structures in which each person is embedded. Thus neither human nor elephant victim of trauma stands alone in their plight. Similar to human violence, an understanding of the aetiology of aberrant elephant behavior, and its amelioration, can only develop when examined in their social and ecological contexts.

The history of the rights and trauma of survivors of war and domestic abuse illustrate this principle. In both cases, individuals were subjected to violence through collective norms and sometimes through law. Both groups have been denied sovereignty over their bodies and limited to identities subordinate to collective purposes—women as helpmates or sexual objects and soldiers as corporal instruments in service of a "greater good."[39] The First World War coincided with the work of Sigmund Freud and Pierre Janet, whom we now might consider psychology's first traumtologists. The Great War became a watershed for both its historical significance and its unprecedented brutality and loss of life. More than five million soldiers died in four years. The conditions of trench warfare, the appalling physical and emotional states of returning soldiers, and the overwhelming numbers of veterans in need of hospitalization or long-term psychiatric care forever dismissed any notions of war as a glorious heroic endeavor. Shell-shock and "neurasthenia," the sanitized terms for mental breakdown, accounted for almost half of British causalities, and the "reality of psychological trauma was forced upon public consciousness."[40]

Soldiers on both sides of the Maginot Line protested against the war. One celebrated case was an antiwar proclamation by the English soldier Siegfried Sassoon. A decorated solder himself, Sassoon was sickened by the pain, injury, and death inflicted upon his comrades merely to suit politicians and a public hungry for victory at whatever cost. "I have seen and endured the sufferings of the troops," Sassoon wrote, "and I can no longer be a party to prolong these sufferings for ends which I believe to be evil and unjust. I am not protesting against the conduct of the war, but against the political errors and insincerities for which the fighting men are being sacrificed."[41] Such a manifesto was grounds for court-martial; through intervention by sympathetic friends, including Robert Graves, Sassoon narrowly escaped the death sentence that the conditions of war dictated for his "crime."

With its numbing horror, the First World War galvanized interest in psychological trauma, and soldiers were seen, however fleetingly, as more than cannon fodder, and the generals and government were seen as something other than glorious martial demigods. The underbelly of war was revealed to be what it was: brutal and anguished. Still, society was not yet prepared to face the realities of war. The thousands upon thousands who returned injured in mind and body failed to break the veil of silence concerning what war really did to people. Experiences of veterans (the disabled were referred to as "moral invalids") continued to be ignored, even belittled. To recognize the emotional ravages of war would be to admit that the government and society at large were agents of trauma. This accusation brought into question the raison d'être of the modern project. Only decades later did Abram Kardiner, an analysand of Sigmund Freud and someone who had witnessed the devastating effects of psychological trauma in veterans, establish a formal psychiatric diagnosis for combat fatigue (another euphemism). Kardiner is credited with coining the diagnosis of posttraumatic stress in his 1941 book *The Traumatic Neuroses of War,* which he wrote specifically to bring recognition and legitimization to combat neuroses.[42]

Soldiers aged, more wars were fought, and the public forgot the devastating legacies of violence, at least for the time being. Public concern submerged in another period of amnesia, and not until the Vietnam War did war-caused trauma begin to get serious attention. Illuminated by such interest groups as Vietnam Veterans Against the War and by the publication of long-term studies on World War II resistance fighters and concentration camp survivors, psychological trauma finally began to move toward mainstream legitimacy within the field of psychology. Along with women's and civil rights groups, civilian antiwar groups also aided the transition. As Herman observes, the link between trauma and politics is critical. Trauma is an "affliction of the powerless," and the "systematic study of psychological trauma . . . depends on the support of a political movement."[43]

Today, *stress* and *trauma* permeate the language and imagery of Western culture, reflecting that "we are onto something widely experienced and intuitively understood"; a wealth of articles and books on the subject can be found in an ever-growing number of fields.[44] The question is how do we reconcile what we know about war, genocide, and suffering and relate it all to elephants? *National Geographic* images of stately elephants walking along the sunset horizon make it hard to believe African elephants can be as stressed out, their habitats as hectic as the traffic-filled freeways of Los Angeles or as terrorizing as Lebanon or New York. Even if nature can sometimes be brutal—"red in tooth and claw," as Tennyson famously put it—the natural world has no analogue to the grinding violence of modern warfare and urban life. Or does it? When we travel to the elephant lands, we find their lives less different from our own than we might think.

4
Deposited in the Bones

I first felt the danger when the birds suddenly became silent and the air stilled. Then came the deep thunder of Man's guns. The cold clarity of memory filled my mind, an image of the massacre of many seasons ago. We lost many, even a few children, too young to be hunted for ivory, but felled in the hail of bullets. Smoke and sound burst from all around, and the calves ran screaming as they saw their mothers fall. I called to the rest to follow, and we ran into the forest, where I knew we would be safe and no Man would find us. We spent many days and nights there, occasional gunfire still heard. We drank water from the river, ate, and rested. The little ones cried, some were too frightened to nurse. I remembered what my mother, a great leader among our people, had taught me. One day I would tell my daughter the story of my family and all that I had learned, and she would tell her children. We would survive.

Upon reflecting on the nature of trauma, Judith Herman writes: "The ordinary response to atrocities is to banish them from consciousness. Certain violations of the social compact are too terrible to utter aloud: this is the meaning of the word unspeakable."[1] The elephant has witnessed such violations. She is known as Min Lyec, the Elephant Matriarch, and her home lies in the Rwenzori Mountains, which extend more than seventy miles along the western boundary of Uganda between Lakes George, Edward, and Albert. The three highest peaks of the range are named after European explorers—or, as the Rwenzori Mountain Enthusiasts refer to them, archcolonists—Sir Samuel Baker, Henry Morton Stanley, and John Hanning Speke, men renowned for exploration that beat a path into the heart of Africa.[2] These are the misty, mythical peaks that Ptolemy called the Mountains of the Moon, whose mysterious flanks and valleys boast a rich diversity of fauna and flora.

Dr. Evelyn Lawino Abe is also Ugandan, an Acholi and elephant ethologist. Elephants are woven into Acholi childhood memories of song, the beat of drums, rich red soil, and fruit of Borassus palms. Abe tells the story of her pregnant mother's encounter with a herd of elephants when Eve herself was yet to be born. Perhaps, she laughs, elephant rumbles to the Acholi are like Mozart to Western babies—that maybe it was the trumpeting elephant matriarch who soothed and tuned Eve to elephant ways as she lay nestled in her mother's womb.

Fate has linked the Acholi people and elephant. Both the elephants and Dr. Abe belong to tribes that have buckled under the heaving violence of appropriation and war since much of Africa was pulled under Europe's possessive umbrella. In both human and elephant communities, adult males were taken away and killed, what wildlife management calls "selective off-take." In both communities, guns were turned onto older females, leading to a precipitous collapse of their respective societies.[3] Yet despite relentless horrors and killings, the elephant and Acholi managed to survive.

Uganda is naturally bisected by the River Nile. To the north is Acholiland, encompassing three districts—Pader, Kitgum, and Gulu. Historically, Acholiland is less populated and more arid than the fertile, developed south. The language is of the Luo, a people who descended from the Upper Nile to Acholiland in the 1500s looking for expansive lands on which to raise cattle and hunt.

Gulu is the place of Abe's birth and where she was raised. Her given name, Lawino, means "beauty and long life," and it is famous from the title of a book by the Ugandan poet Okot p'Bitek, *Song of Lawino*. P'Bitek's Lawino is portrayed as the defender of the culture, but to see culture as something solely human is shortsighted. The elephant is the Acholi totem. The Acholi people never eat elephant meat, Abe says, because they believe that the very existence of their lineage depends on the elephant.

Abe has lived many years in forced exile. Hers is an unreal life, suspended between two lands, between two lives; she is a child of Ugandan soil living unclaimed in another country. "I had heard of the movie *Alien,* and it captivated my imagination. I related them to unidentified flying objects—subjects of contention, things that leave earth people bewildered and are linked to plane disappearances and symmetrical circles in fields of crops. Now I too belonged to that category. To abandon one's culture and learn a new one and become labeled as an Alien or Refugee."[4]

Abe completed her doctorate at the University of Cambridge in 1994. The title of her dissertation—"The Behavioural Ecology of Elephant Survivors in Queen Elizabeth National Park (QENP), Uganda"—promises something remarkable, hinging on one word: *survivor.*[5] *Survivor* is a common enough term, but it is generally reserved to describe humans, humans who have lived through affliction. Abe studied elephants, not people, and the effects of large-scale killing from poaching that was inflicted on their communities in Uganda's flagship national park, the Queen Elizabeth. In the 1960s Uganda boasted an estimated thirty thousand elephants. By 1982 there were a mere two thousand survivors.

A more scientifically conventional title might have read "The Effects of Poaching on Elephant Behavioural Ecology in QENP," or "The Effects of Large-Scale Mortality on Elephant Behavioural Ecology in QENP"; elephants and other animals are typically discussed as objects, upon whom things and people act to produce a programmed response. Abe's use of *survivor* implies agency, selfhood, and righteousness in elephants' persistence, resonant with what we have learned from Happy and the other elephants.

Survivors are individuals who weren't killed but were left stranded on cultural islands after floods of violence receded. *Survivor* carries other reminders. The term is most commonly associated with human traumas: the European Holocaust, the Rwandan genocide, the systematic elimination and subjugation of American Indians, and other iconic atrocities of the past centuries. Elephant survivorship invokes these ghosts in the same breath, the chosen few who are joined through a common effort to hold the shreds of their respective cultures together.

None of this narrative is explicit in Abe's dissertation. Each page reflects scientific rigor and is served with the flavor of clipped British precision and reserve. Yet something else, something rare, emerges, a sense of the subject and the writer that transcends impersonal facts and figures. It comes from Abe herself, shining through scientific architecture and speaking of a world beyond: a way of life that pulsed through elephant lands for thousands of years. It is a world far away from the reductive boxes of Cartesian thinking, a way of life that does not cut through ancient ties and shared, inherited roots.

The ancestors of Min Lyec and our ill-fated young South African elephants first settled on the continent more than eighty million years ago, and they evolved into their present modern physique between five and six million years ago, at roughly the same time that our human forebears were living in today's Kenya and Chad. Humans and elephants have shared a long history, living side by side in relative equanimity. If they had not, European colonists would not have seen the vast herds they did upon arrival. The particular difficulties besetting people and wildlife today are new to the history of the two species.

Whether as targets for gun or for camera, African wildlife has been a draw from the beginning of colonization, and the elephant a particular favorite as target and victim. While the trunk may be the elephants' calling card, another part of their anatomy has brought calamity to the species: majestic ivory tusks. "White gold" entices hunters from all over the world to both continents. Along with trophy hunting and the inexorable push by humans into ancient elephant lands, the hunger to possess ivory has brought elephant culture to its knees.

Impacts on elephants from abroad date back thousands of years. In the first century C.E., Pliny the Elder, author of the *Naturalis historia* that has been used by scholars to study the plants and animals of the natural world for centuries, complained about the shortage of ivory. At the turn of the first millennium, ivory was relatively hard to find because the abundant herds of northern Africa had already been decimated by hunters. As time went on and elephant numbers continued to drop, the ivory trade moved southward, taking down herd after herd.

In 1652 the cape was colonized and Europeans "introduced both a strong market economy and firearms, starting the overexploitation of wildlife." With colonization came an entire belief system that contrasted dramatically with that of the indigenous peoples: "Christianity excluded beliefs in the intrinsic power and value of nature, as believed by hunter-gatherers, and commanded its followers to tame and civilize nature in the service of humankind."[6] South Africa's interior remained somewhat resistant until the nineteenth century, when recreational hunters established settlements in places where people and elephants had lived in relative peace. It has been estimated that in 1855 ninety thousand kilograms of ivory was exported from the Transvaal.[7] At the same time that wildlife suffered huge declines, so did African peoples. As elephants were felled by European guns, tribes of present-day South Africa—KwaZulu, Venda, San Bushmen, and many others—went a similar way.

In a parallel to North American occupation, European-backed hunting was instrumental in bringing about radical changes to African cultures through displacement, slavery, and destruction of the indigenous peoples' livelihoods. Magqubu Ntombela of the Wilderness Leadership School speaks of the land and traditions before colonization:

> KwaZulu was once a land full of wild animals like the elephant, rhino, kudu and crocodiles. We lived with and knew these animals. I was born amongst them. This animal is highly respected by our people. . . . We did not kill the animals

> without permission from our traditional king, King Dinizulu. He did not allow people to kill the animals and any person caught was severely punished. . . . I think that it is a very good thing that we should stick to the old traditional ways of living so as to protect the future for our children, so that our children will understand what a wild animal is. . . . I understand the plants and the animals, birds and insects. I can tell when the rain is coming. All this knowledge is in my blood. . . . We once had a way of living in the world and knowing what was happening on the land. We were in tune with all that lived and sang.[8]

As tribes—both wildlife and people—were pushed out of their homelands, the colonizers took hold and replaced traditional practices and values with the necessities of survival in a Europeanized culture and economy. Some indigenous peoples acquired firearms, which they used as valuable bartering items and to fight the settlers. Eventually, African wildlife was beset by weaponry fired by both Africans and Europeans, and the burgeoning power of industrialized markets in nineteenth-century Europe and the United States fueled the ivory industry.

The scale of South African elephant killing in the 1800s is astounding. One Henry Hartley is said to have killed more than 1,000 elephants in his career, and on one expedition alone Jan Viljoen shot 210 elephants.[9] Exploits of big-game hunters tantalized the imaginations of new waves of pioneers and adventurers to Africa—Ernest Hemingway, Isak Dinesen, and scores of others. In the 1920s George Adamson, who later became a great conservationist, and his brother occasionally purchased an elephant license for a mere twenty-five pounds, and with heavy rifles could recover one hundred pounds for the tusks they brought in. The film *Out of Africa* vividly captures the breathtaking beauty of the continent but belies the slaughter that accompanied the footsteps of colonialism. Once stretching across the continent, elephant migration has largely ceased through the fragmentation of habitat and the disintegration of elephant culture. Towns, roads, and fences crisscross ancient elephant highways, and lands converted into agricultural fields are guarded by watchful villagers. And then there is war. Uganda's story is one of many variations on the theme across continents.

In 1877 Anglican Christians came to Uganda and began systematic colonization. By 1894 the British established a protectorate that was centered around the southern kingdom of Buganda, the site of today's capital, Kampala. During occupation, economics and politics shaped the map: the south received the bulk of development, while the north provided most of the labor and military manpower. This differential was soon to play out in hostility.

Independence from Britain came in the late 1950s, and civil war followed soon after. Within four years, a half million people fled the country, and as many were murdered. Successive political battles ensued, and Uganda slid into nearly a half-century of violence, which continues today. The Acholi have been among the victims of that chaotic regime. As has been true in other nations beset with civil war—Rwanda, the former Yugoslavia, and Haiti, for example—killing seems to stop only when the people have been bled dry.

Beyond the losses from slavery and colonial suppression, Uganda has suffered two officially recognized genocides. The first, which occurred from 1971 to 1979, during the rule of the infamous General Idi Amin, was depicted in the recent film *The Last King of Scotland*. Even after he was ousted, and Milton Obote was reinstalled as president in 1980, Uganda's civil war, the "war of the bush," continued with a second wave of fighting and deaths. During "Obote II," entire communities were vilified, and more than a quarter of a million people in northwestern Uganda escaped to Sudan and Zaire (Congo). By 1985, when Obote was removed from office by a military coup, more than half a million were dead.

When Yoweri Museveni took power as president in 1986, Uganda had new hope. Kampala was one place that profited from the change; the city now has a relatively strong economy and a reputation as one of Africa's safest cities. But across the Nile to the north, there is no sense that war has ended. Today's war in Acholiland is ostensibly between the Museveni government's National Resistance Movement (NRM) and an opposition whose leadership and composition has metamorphosed over the years. A succession of spirit medium leaders have led the resistance, their express mission to "cleanse the Acholi from past sins." Alice Auma (Lakwena) was one, and she organized the Holy Mobile Forces. Within a year, Auma suddenly lost her spiritual powers, and Joseph Kony's Lord's Resistance Army (LRA) became the most powerful opposition to the government. In January 1997 LRA soldiers burned and looted thousands of homes in the north and mutilated and killed more than four hundred men, women, and children. Only vestiges of community and family structure are left; most Acholi have left the country, died, or settled semipermanently in camps.[10]

An entire generation has been raised in camps for internally displaced persons (IDPs), living behind wire fences and facing daily violence. Some venture out occasionally to visit their former homes, but the risk of capture, torture, or death from marauding soldiers is high. Unlike Auma and others who preceded Joseph Kony, the LRA directs most of its attacks and violence against civilians. Women and children are brutalized and tortured. In the past, Acholi parents warned their naughty children about Ojok Gacci, an Acholi "bogeyman." He came to life in the form of the LRA, whose soldiers abduct children as young as five and six to be slaves, soldiers, or prostitutes.

To ensure their loyalty, the children are subjected to gruesome rites of passage; a soldier-to-be, for example, may be forced to kill his own sibling or torture someone. One boy, Jack, describes his abduction experience: "I was taken in 1995, when I was ten. It was at night when we were asleep. After we'd been caught we were tied up like slaves. We were taken to Sudan for training. I saw many children killed. They are killed with a machete to the head. It happened all the time. They make children kill other children. I had to kill other children or they would have killed me."[11]

Girls are routinely raped, beaten, and burned. Sarah, another abducted child, a young girl, talks about her abduction:

> I was taken in 1994 when I was 13. I was at home in Wol. Seven of us were taken, four boys and three girls. I was with

> the rebels for two years. After about a week I became the wife of a soldier. That's what happens to the girls. We went to Sudan. We were trained, and eventually I was given a gun and uniform. Kony comes to address troops returning to Uganda. The young ones are supposed to lower their eyes, not to look at him. You have to obey the Spirit's command. For example, the Spirit may say: "You must not kill any chicken for the next three weeks." Then after three weeks, when you kill a chicken, then you can start killing people.[12]

Most children do not return to their homes even if they are able to escape. Often, their parents have left or are dead. To escape abduction and camp brutality, thousands of children become "night commuters," taking to the roads and byways, carrying bundles wrapped under their arms, to roam en masse, hoping to find a place safe enough for sleep.[13]

Mass deaths, destruction, and the exodus to other countries have left northern Uganda bereft of elders, who made up the traditional infrastructure that defined Acholi life. Uganda has also become a victim of "slim," as HIV-AIDS is called because of its wasting effects. The virus is the number one cause of death in Africa and has contributed to the growing number of parentless children who must live alone and untended in the camps.[14]

Loss of father, mother, and community means more than biological vulnerability: it undermines the sense of self and deprives a child of the cultural and psychological wherewithal for survival. Northern Uganda is without what an elephant ethologist would call socioecological knowledge, the wisdom and tutelage of mothers, fathers, grandparents, and the intact communal network that nurtured young Ugandans into maturity. Now the stories, names of healing plants, and history exist only in vestiges of the few remaining elders' memories. The Acholi speak sorrowfully of the tatters of a rich, proud culture, one that is composed of inexperienced young men and women. In place of traditions, a new mythology has grown from seeds of strife. Without a common restorative narrative as an antidote to fear and pain, there is no guiding path that might lead to a new future, away from the void created by violence.[15]

What has happened to the Acholi has happened to their totem. Poachers aim their AK-47s on elephants and follow with machetes to hack off the ivory without waiting until the animal is dead. According to Abe's research, by 1994 elephant populations had been reduced by 90 percent, leaving the herds skewed in age. At QENP alone, 30 percent of the 273 elephants were under five years of age, and only 28 percent of the males were over thirty; effects of these demographics are reflected in DNA analyses. Motherless bands of young elephants, sometimes in the hundreds, wander looking for food and protection. At times, they come into the camps or deserted farms to flee from the smell of cordite and booming guns.[16]

The decline of the elephant family is an often-repeated story. In North Luangwa National Park, Zambia, 93 percent of the population has been killed, and traditional herds composed of mothers and allomothers are virtually absent. Many infants are reared by inexperienced, highly stressed single mothers without the detailed knowledge of local plant ecology, leadership, and support that a matriarch and allomothers provide. According to the studies in Zambia by Mark and Delia Owens, mothers are younger than in the past: 48 percent of births were to females less than fourteen years old, compared with a normative mean age at first birth of sixteen years. Thirty-six percent of groups have no adult females, one-quarter of the units consist only of a single mother and a calf, and 7 percent of groups are made up entirely of sexually immature orphans.[17] In Mikumi, Tanzania, the population has been similarly affected, and in Uganda, elephants live in semipermanent aggregations of more than 170 animals, with many females between the ages of fifteen and twenty-five years but no traditional familial association or hierarchical structure.[18]

Elephants must exist like expatriates in their own land, lacking the meager protection that refugee camps can sometimes afford to their human counterparts. Parks, which in some ways resemble IDP camps, fail to offer sanctuary from marauding soldiers and villagers. Even the landscape is not immune. An African saying, usually taken metaphorically, today has literal significance: "When two bulls fight, the grass suffers." Elephant declines have altered the landscape from grassland to bumpy and thorny thickets, so much so that remaining elephants are forced to seek out suitable browse and water in areas that bring them into further conflict with people. IDP returnees find starving wildebeest and elephants moving through fields of millet, maize, and sweet potatoes of once-abandoned villages. In the Lango subregion, more than four hundred families have been displaced by "marauding rhinos and elephants," prompting a local leader, John Bosco Okullo of the Amuru district, to threaten: "We shall organize the community to send back the elephants if the concerned authorities fail."[19] "Elephant attacks" have also started. The numbers are not high, but there is a marked change in elephant behavior, a sense of increased aggression towards humans, and some human fatalities.

Where once elephants were able to secure protection, life has become as dangerous as on the outside. The Zimbabwe army is killing and eating elephants, and in the Congo, Virunga National Park director Emmanuel de Merode announced, "We've definitely lost 20 percent of the [elephant] population this year and probably more" as a result of poaching spurred on by China's ivory lust.[20]

Events of recent decades are merely the worst of a long history of human-caused starvation, hunting, mass culls, and poaching that have reduced elephant numbers in Africa from an estimated ten million to a few hundred thousand. Incursions into elephant lands continue as human population and consumption explode. Kenya's population has jumped from 8.6 million in 1962 to more than 30 million in 2004, and between 1973 and 1989 elephant numbers plummeted from 167,000 to 16,000. Furthermore, because elephants are increasingly isolated on habitat islands dotted across the waters of human occupation, they are under constant stress to find food, water, and space, while facing attendant hostility from farmers whose crops get trampled and eaten.[21] These impacts are even believed to have contributed to genetic, behavioral and physiognomic change.[22] In 2002, out of the 174 elephants at Addo, where such high male-on-male mortality has been observed, 98 percent of the females were without tusks.

Asian elephants have had a different history from that of their African counterparts. Elephant worship is as old as Indian culture itself. Ganesha, the elephant deity, is revered for bringing fortune and blessings. In Kerala, India, "people adore captive elephants as an incarnation of Lord Ganesh, the elephant-headed God. It is one of the best loved animals and an integral part of the socio-cultural and traditional functions of Kerala. The very sound of the clanking of its chain attracts people irrespective of caste, creed, sex, and age to its side in common fascination."[23] This elevated status has brought protection to the living pachyderm in ways not shared by its counterparts in Africa. Until recently, elephants were rarely killed in India. Yet despite their divine status, the lives of many have not been easy: elephants have been domesticated for millennia.[24]

Elephants who are brought into service are either wild-caught or born in elephant camps. The capture and breaking of an elephant is called *mela shikar.* Riding trained elephants *(kunkis),* the mahout (trainer and handler) captures a wild elephant by slipping a noose around his or her neck. To avoid strangulation, the elephant finally stops fighting to get loose. More knotted ropes are tied onto the neck, the back legs are tied together, and so are the forelegs. If an animal is particularly difficult to control, his head is tied to the ground such that he cannot move without being strangled. Water and food are withheld to help break the elephant down. Eventually, the elephant is subdued, chained, and taught commands. Finally, when the mahout has broken the animal, "the transformation of a wild creature to a gentle captive elephant becomes the sign of another glorious interaction between the elephant and man. A domesticated elephant is born."[25]

Elke Riesterer, who describes herself as a "masseuse for all species," witnessed torturous treatment of a bull elephant in musth. One day, while visiting a mahout camp associated with a veterinary clinic, Riesterer, with several veterinary students,

> came upon a bull chained at the end of a shallow pond next to a hillside. He pulled desperately at his chain and rocked back and forth. An elephant in musth will lose much money for his

owner since the bull is unable to work. In order to shorten the length of musth, which can last anywhere from a few days to months, the owner tied the bull to a tree on a very short length of chain. He had stopped feeding the bull and provided very little water. The veterinarian students clamored around as closely as they dared to witness the spectacle, curious yet detached from the bull's tortured experience. Agitated, the bull yanked at his chain, trying to pull farther away from the crowd. The fluid still streamed from his temples. Tears welled up in my eyes and it made me realize how callous humans can be to torture a creature like that.[26]

The bullhook, or ankus, has been standard equipment in training for domestication. It is a wooden pole with a curved metal hook at one end used to inflict pain on sensitive points, including the genitals, mouth, and anus of an elephant being broken. According to Indian veterinarians, ancient texts, the Gajahastra and Nilahastra, describe these sensitive spots—somewhat analogous to acupuncture points—as elements of a "special language" by which to communicate with and control the elephant.[27]

The mahout profession was modeled on the royal dynasties; elephant "kingships" were passed from generation to generation, preserving the accumulation of detailed information of elephant behavior and care. In the past, mahouts and their elephants lived close together, even sleeping near each other, and these relationships often lasted a lifetime. However, with modernization and replacement of local economies by global ones, traditional practices, including elephant training, have been disrupted.

Today, the costs of keeping elephants far exceed the perceived benefits of mahout training, and mahouts no longer hold a respected station in society. Would-be mahout sons leave for the city looking for easier and safer jobs than those of their fathers. In their place come untutored young men who have no other opportunity to make a living. They are typically unfamiliar with elephants, having spent little time with their animals, and they lack allegiance to their profession. In the past, being a mahout was a lifetime career, but over the past few years, the average tenure has dwindled to about twelve years. Young men begin as mahouts after only one week training. As a result, cultural knowledge about elephant psychology and biology is replaced entirely by brute force, as the mahout arms himself with pronged iron ankus, knives, and sticks and employs crude methods of confinement in order to control a large bull elephant.

Peter Jaeggi, a Swiss journalist who has followed the plight of Asian elephants for the past decade, interviewed a mahout trainer about how he selects new trainees: "I pick out the poorest people; young, homeless people who can neither read nor write and therefore have no chance of getting any other type of work."[28] In Sri Lanka, half of the mahouts consume more than one bottle of alcohol a day. Violence against elephants has increased, and so has the number of mahouts who are injured and die. Several hundred mahouts are killed annually in Asia, where such deaths used to be fairly rare. Elephants who kill their mahouts sometimes not only crush the keeper but tear and toss the human fragments far and wide. Indian historians and older mahouts attribute the increase of elephant attacks to the abandonment of traditional mahout training methods and the diminished quality of the mahout-elephant relationship.[29]

When not used for ceremonies at temple celebrations and weddings or to beg in the streets—a practice that has become more common where elephants no longer have "employment" and their owners demand that their charges earn their keep from tourists—Asian elephants are often subdued with opium or other drugs. Sometimes they literally die on their feet from exhaustion. Elephants used for logging are drugged with stimulants; when they are not working, they are usually chained. The stereotypical behavior commonly observed in elephants in captivity of endless bobbing and swaying has never been reported in the wild. Fredrick Kurt, a noted elephant researcher of Asian elephants in zoos, submits that "weaving elephants are animals that went insane in the chained imprisonment."[30] And there are new problems. Scores of Thai mahouts bring their elephants to Bangkok, seeking to eke out a meager living. Ninety percent of their elephants come from Surin province, hundreds of miles away. Twenty years ago, the Thai government banned logging there, putting the mahouts and their elephants out of work. In the city the mahouts sell rides or coax tourists to buy sugar cane to feed the elephants.[31] This practice is lucrative enough to stimulate the illegal taking of wild elephants, which further undermines a dwindling population estimated at between two thousand and twenty-five hundred. Elephants are often trapped in deep pits, then beaten for three solid days. Pang Kanjana is one unfortunate example, an adult pregnant female elephant who is crippled with a deformed left rear leg, which was broken in her capture and healed improperly.[32]

Iain Douglas-Hamilton, who has studied African elephants for more than forty years, spoke about the change in attitudes toward Asian elephants: "Inflammatory words are used daily about elephants: terrorize, are a menace, are on a rampage, running amok, or are rogue"; elephants, he said, "are weekly being shot, snared, electrocuted, run into by trains, poisoned in retaliation, and everywhere deprived of habitat."[33] Dr. K. C. Panicker, an elephant veterinarian in India, documented the types of elephant deaths he and his clinic alone have encountered from 1976 to 2003:

Death during work: 18
Death after veterinary procedure: 4
Death after torturing and shock induced by mahouts and other people: 50
Death after blood poisoning caused by torturing as well as arthritis: 40
Death caused by undernourishment and too strenuous work: 50
Death after being shot by police: 10
Death after accident with motor vehicles: 13
Death after constipation caused by inappropriate diet: 40
Death after heart attack: 15
Death caused by electrocution: 3

Death after explosion of an explosive device in the elephant's mouth: 3
Death by drowning in a water tank: 2[34]

"Considered an asset in the past," the elephant "is now becoming a liability." Armed with torches, clubs, and sometimes guns, vigilantes hunt down local elephants. Nor does the elephant's traditional status as deity carry any weight. Pastel renderings of Ganesha on the walls of Calcutta and Delhi have the name of Osama bin Laden scrawled across them, and newspapers are filled with stories of villagers battling elephants. (Indeed, one elephant in India dubbed Osama, said to have killed eleven people, was caught and killed in Jharkhand, shot twenty times.)[35] Similar to African elephants, Asian elephant numbers have plummeted, but exact numbers are uncertain. Thailand's permanent secretary for natural resources and the environment, Saksit Tridech, states, "We expect captive elephants to disappear within the next 14 years, which means wild elephants will again be under threat from hunters to serve the high demand in the market."[36]

Human violence, elephant violence. The scholar of postcolonialism Frantz Fanon once said that the deleterious effects of oppressors were left "deposited in the bones" of their victims. According to one theory, homicide represents a sort of "conflict assay," an index of the health and status of the social contract.[37] The same seems as true of pachycide. Injurious crime is more than an act of aggression; it is also a relational breach and a rending of social fabric that, as we have learned from neuroscience, leaves scars on the psyche as well as on the body.

If we were to draw a map using attachment theory and traumatology as guides to trace the trail of violence from "neurons to neighborhoods," we might do well to begin with the dramatic disruption of indigenous cultures by colonialization—a disruption passed through successive generations and transmitted across species from human to elephant and back again to human, like a virus. Traumatic disruption takes on a life of its own, becomes a culturgen, redrawing a perceptual and moral baseline to define culture itself. Cultural trauma replaces traditional ritual and leaves "indelible marks upon their group consciousness, marking their memories forever, and changing their future identity in fundamental and irrevocable ways."[38]

When we look through science's dark glass, the human-elephant conflict takes on a whole different meaning: aggression between the two species is not homegrown violence but is related through time and space with the disruption of ancient social bonds. A pebble thrown into the water, violence ripples outward in increasingly bigger circles, eventually exploding into the animal kingdom at large. Scientists assessing African human-elephants relationships concede that human impact before European colonization

> was sufficiently minimal that the wildlife persisted in the presence of humans for millennia, and thus the use of African fauna as a food source was sustainable most of the time. If not, we would not have had reports from early European explorers describing Africa as a place "teeming with wildlife." Although contemporary academic theories that develop traditional African values about the environment into theories of environmental ethics are scarce in South Africa, sufficient clues exist that enable a partial reconstruction of such values. Perhaps the most important fact to consider is that African people lived alongside wildlife for centuries without hunting any species of wildlife into extinction that we know of, as has happened on other continents like Europe.[39]

How to retrieve history? Uganda shows signs of recovery. Camp residents have started to return to their former homes and businesses, and Kampala has acquired an air of energy and viability. But camp returnees now find themselves in competition with their erstwhile elephant neighbors, who are also in flux from the turmoil. And there are plans, backed by outside investors, to build an eighteen-hole golf course in Queen Elizabeth National Park, just one example of outside monies targeted to revitalizing a lucrative tourist trade.[40] The golf course is to be located on elephant migration routes and will consume what scant habitat is left. Plans to build a major interstate highway threaten India's largest surviving elephant population. In Sri Lanka there is talk of capturing the entire wild elephant population and domesticating them, subjecting them to brutal training and a life in servitude, or what one scientist calls elephant "ghettos," parks that include only a fraction of natural home range.[41] The Coordinator for the World Wildlife Fund (WWF) Asian elephant and rhinoceros program, Christy Williams, warns that the "billions of dollars lined up for regional and national level infrastructural investments such as the Trans-Asian highway project and various hydro-power and irrigation projects are going to significantly increase human-elephant conflict across Asia."[42] The condition of the native elephant is now a nervous condition.[43]

What is the future for a culture whose totem is the elephant? What is the future for the totem's own culture? Recovery from bone-penetrating violence involves more than a cessation of war and economic revival. It requires embarking on a path different from the one that brought the violence. As neuroscience and the altered behavior of so many elephants suggest, changes to ancient patterns governing elephant habitat, social systems, and beliefs have affected the building blocks of life itself. The traces of chemical changes and alternative neuroendocrinal pathways that took place hundreds of years ago have not been erased: they linger in the minds and lives of people and elephants who live today.

5 Bad Boyz

Suddenly one day, while he suckled from his mother, there was an explosion. Noise filled the air, dust blew up in clouds, and the earth trembled. He could hear his mother and aunts calling frantically, but he lost sight of them in the confusion. When he ran to find them, he was caught in long, painful thorns. The next thing he remembers is being grabbed and tethered with coarse rope to the warm, motionless bulk of his mother. When the shooting stopped, he was pushed inside a truck with others, and after many hours, they arrived somewhere he had never been. There he met other elephants, but no typical family to shelter and reassure him. Food was hard to find—he had difficulty recognizing what plants were safe to eat, and so he often went hungry. Even the air was different from home, and frightening sounds haunted the nights. It did not seem possible to overcome his grief. He felt an overwhelming sense

of homelessness and hopelessness. But day by day he learned how, if not to live, to survive.

A peek back into history and into current science tells us that elephant violence has less to do with elephants than with our own culture. Elephant attacks on rhinoceroses and mahouts reflect the violence that this otherwise peaceful species has experienced. Hyperaggression, depression, infant neglect, and other symptoms are not uncommon for elephants in captivity, but are unheard of in the wild.[1] When we know elephants' history and their sensitivity to surroundings, behavioral oddities cease to be a mystery. Elephants are merely mirroring the circumstances in which they have come to live.

Elephant behavior also testifies to the powerful effects that humans have as agents of natural selection. A study based on forty species has found that "organisms subject to consistent and strong 'harvest selection' by fishers, hunters, and plant harvesters may be expected to show particularly rapid and dramatic changes in phenotype," and in fact, "harvested organisms show some of the most abrupt trait changes ever observed in wild populations, providing a new appreciation for how fast phenotypes are capable of changing."[2] Rates of evolutionary change are three times as high in species that have been selectively harvested—that is, those species whose individuals are preferentially killed for their size, antlers, tusks, or other characteristics: a trend that is likely to continue as wildlife become increasingly contained in small, isolated populations.[3]

Under such conditions, human-elephant conflict (HEC) takes on a very different meaning. As the wildlife veterinarian Mel Richardson puts it, issues surrounding elephants are "not about the animals."[4] Rather, they are about humans: human-elephant conflict revolves around questions of social justice and human introspection. Much like other cultures that have refused to be absorbed by colonialism, elephants are struggling to survive as an intact society, to retain their elephant-ness, and to resist becoming what modern humanity has tried to make them—passive objects in zoos, circuses, and safari rides, romantic decorations dotting the landscape for eager eyes peering from Land Rovers, or data to tantalize our minds and stock the bank of knowledge. Elephants are, as Archbishop Desmond Tutu wrote about black South Africans living under apartheid, simply asking to live in the land of their birth, where their dignity is acknowledged and respected.[5]

As part of this new trans-species framing, we move from a purely ethological treatment of aberrant elephant behavior, the approach conventionally taken to study elephants, to one of psychology, whose theories and methods were developed to address matters of the human mind. In merging ethology with psychology we recognize what humans and elephants share in brain and behavior, and we learn to ask more expansive questions about elephants than are usually included in ethological investigation. For example, given what elephants have experienced, who is the elephant of today: not just what they do but *who* they are? Who is the young, orphaned elephant mother now listless and raising her infant with maternal indifference? Who is the bull in captivity who turns on the older group matriarch and kills her?[6] To answer these questions, to explore the effects of trauma, we must delve more deeply into understanding the elephant mind and society, as an elephant social psychology begins to take shape.

Ethology and psychology overlap but are not synonymous. In ethology, behavior is everything. In psychology, behavior is only a part of the bigger picture that includes details about thinking, feeling, and states of mind. For example, a certain behavior, like picking up a knife, may be stimulated by very different sets of mental states and motivations. A man standing in his kitchen holding a knife could be feeling quite calm, ready to make dinner, intending only to sever the tips of fresh summer green beans. But should a stranger burst into the room shouting and holding a gun, our man might hold the knife with a wholly different intention. Clearly, behavior does not always map exactly to mental states, even in the matter of killing. People kill for an almost infinite number of reasons: jealously, fear, material gain, revenge, or simply by accident—each theoretically governed by a different motivation and emotion. Small wonder that attorneys go to great pains to establish the mental state of the accused. A classic example is illustrated in the case of Dan White.

On November 27, 1978, Dan White, who had served in Vietnam and was a former policeman and city supervisor, shot and killed San Francisco city supervisor Harvey Milk and mayor George Moscone. Prosecutors, pressing for a murder conviction, argued that the killings were premeditated, that White had held the two men responsible for derailing his political career. An expert defense witness, however, a psychiatrist, testified that White had showed "diminished capacity" and therefore was not entirely responsible for his actions. The psychiatrist claimed a diet change to junk foods, including Coca-Cola and Hostess Twinkies, could have exacerbated White's mood swings. There was no disagreement about White's actions, but they were ascribed to different motivations and mental states. The so-called Twinkie defense prevailed: the court ruled that White did not have the mental capacity at the time to kill with premeditation.

Knowing how difficult it is to ascribe motivation and mental states to humans, we face a formidable task in understanding what the young bull elephants were thinking when they copulated with and killed rhinoceroses. What was on their minds? Hard question, but insights and at least some of the answers lie in their personal stories.

Southern Africa, comprising Angola, Botswana, Malawi, Mozambique, Namibia, South Africa, Swaziland, Zambia, and Zimbabwe, has the largest range for elephants of any region on the continent and 39 percent of the species' population. South Africa accounts for 2 percent of this regional range, or 0.8 percent of the continental total. There are about thirty-five individual elephant populations in the country, and an estimated 17,000 animals overall. Kruger National Park has by far the largest single population, approximately 13,000 (14,500 when neighboring private reserves sharing an unfenced boundary are included).[7]

Kruger National Park is the wildlife flagship of South Africa and, some claim, of all of Africa. Roughly eight thousand square miles, it hosts an amazing array of birds, mammals, reptiles, and landscapes. Imperial ground hornbills move through waving grass like navy ships, giraffes glide through treetops, and the cry of raptors echoes in brilliant skies. Not only is there tremendous faunal diversity, but the numbers of wildlife hark back to earlier times before the disappearance of wildlife outside protective reserves.

In 1926 Kruger National Park came into being through National Parks Act no. 56. Like most parks around the world, Kruger was established in response to catastrophic declines in wildlife. Before this, the legislature in South Africa had enacted laws seeking to curb the scale of hunting, but they were no match for the prevailing pioneer mentality, one that persists and complicates the task of elephant conservation. As in many countries, including the United States, hunting and commercial use of wildlife are ingrained in South African society, even to the extent that "sustainable use" has been written into the democratized South African constitution.[8] South African parks reflect strong conservation values but again, as elsewhere, are financed by tourism and hunting revenues. In reality, wildlife safety is provisional and protection is determined by regulatory interpretation.

In 1965 park officials decided that Kruger's elephants were overprotected and that their numbers needed to be kept at a steady 7,000 individuals. From 1966 to 1994, 16,201 elephants were culled (that is, killed) or removed.[9] The rationale for this action remains controversial, and as we shall see, the policy continues to threaten elephant life. Kruger culls have been a primary cause of the orphaning of young elephants.

In Rhodesia, later Zimbabwe, national parks authorities culled some 50,000 elephants between 1966 and 1996, a large number of them going to the United States as trophies, evidence of the significant influence of international interests. In his book *The Fate of the Elephant,* Douglas Chadwick interviewed a former Zimbabwean National Parks game ranger named Adrian (Adie) Read, who worked on elephant culling teams during the 1970s and 1980s. Chadwick describes Read as a "professional elephant terminator," "an athletic man in his thirties, [who] ranched game, farmed crocodiles, and ran safaris." Read provides a first-hand description of what the South African young bulls might have experienced:

> We had one light airplane—a Super Cub. There were three hunters: a center, who is the boss man, and two flankers. Each guy had a radio for communicating with the others and with the scout plane. Each shooter had a gun bearer, too. We used .458 and .308 [7.62 mm NATO caliber] semiautomatics. The bearer carried a second weapon and spare ammo. It was his job to reload while the hunter shot. Behind us came the trucks and 250 laborers for skinning and butchering.
>
> On a normal day, we were up at 5.00 A.M. and the pilot got up into the air soon after. If he saw a herd of more than forty, he called in, and we got ready to go to work. Less than that wasn't economical. Say the pilot saw fifty. Off we go before the day starts heating up. The pilot tells us the wind direction over the elephants. We stop our vehicles two kilometers away and walk in through the bush. Vultures learned to follow the plane, and we had to abort the mission now and again because the vultures were too thick to fly.
>
> The whole strategy was to be deployed so as to be able to take out every single elephant. So the outlying ones have to go right off. The man in the center fires first, usually at one on the outside. People seem to think we start with the matriarch so others will stay with her once she's down. No. Often we wouldn't see her until halfway through. But if you screw up or she gets your wind, she'll come charging, and you have to drop her quickly. The matriarch is always very big and very cheeky.
>
> With experienced guys, the actual killing part is incredibly quick. We got ninety-eight dead elephants in under a minute with three shooters once. If they don't bolt, you can do that. You have four rounds in the chamber. You fire three and give

> your gun to the bearer to reload—if you know what you are doing. Never all four. Otherwise, you turn and your bearer is gone, panicked, and you're left with an empty gun facing an elephant charge. You shoot only in the head. The animal is dead before it hits the ground. If it is running away, you have to hit the spine, then run around and shoot it in the head. Ninety percent of ours were shot in the brain.[10]

In listening to Read's narrative, Chadwick comments, "I was still trying to imagine ninety-eight giants felled in less than sixty seconds. That was superb shooting, I told Adie. I had no idea such a thing was possible. It took scarcely longer than blowing them up with a bomb."

Read continued his description: "In the years I was culling we got, oh, around 15 000 elephants." Chadwick interjects that at the time, this number was "roughly the number left in Kenya." Read's story once again resumes as he details the cull event. "I can safely say the *one* wounded elephant got away, up around Lake Kariba. That's it. Now the strange part: If we went culling one day, we could go out the day after and shoot in the same area, because nearby elephants come over to investigate. No doubt about it, the message gets to other herds, even if you've killed every one in the first group. It's that infrasound. It has to be. . . . You go after a wounded calf, and even though the pilot hasn't seen another group anywhere close, the calf will run to the nearest herd. In fact, you get this sort of thing . . ." Read then "grabbed some paper and sketched the route of a fleeing baby elephant. The pencil started off racing aimlessly, headed more or less in the opposite direction from the large X signifying the closest live herd, ten kilometers distant. But gradually and unerringly, the pencil arced around until it was headed straight toward that faraway herd. Infrasound." The episode closes with Read saying that if the baby elephant made it to the other herd, then "we'd often leave it alive." Many of the calves captured in Read's and similar culls went to the United States and abroad, while others formed the foundational populations for diverse parks and reserves.[11]

In the 1960s wildlife translocations were taking place, but it was in the 1980s that a significant number of animals began to be shuffled around South Africa to various privately and publicly owned reserves, including Pilanesberg National Park.[12] Because of its proximity to both Johannesburg and Pretoria, with their dense human populations, the area surrounding Pilanesberg has lost much of its wildlife. The young bull elephants were some of the first to be seen in many years. Pilanesberg's 550–square kilometer size is on the small side, but its perimeter encircles a marvelous landscape bisected by ragged hills marbled with acacia and other bushveld species.

Elephant home ranges—the average extent an elephant traverses to forage, socially interact, mate, and live—can vary tremendously. For example, the average home-range dimensions for an elephant female group has been observed to range from 240 square kilometers in Kruger to 1,800 in Tsavo. Relative to other areas in southern Africa, elephant home ranges in South Africa tend to be smaller because the populations are kept fenced in conservation areas and reserves: the mean for breeding herds is estimated at 595 square kilometers and for bulls 153 square kilometers; throughout other parts of southern Africa the average breeding herd home range has been measured at 1,678 square kilometers, 2,095 square kilometers for bulls.[13] On average, an elephant walks between 8 and 20 kilometers a day, although this varies with season and circumstance. In the arid region of Etosha, elephants sometimes stay on the move for eighteen hours at a stretch and can travel up to 180 kilometers in a day.[14] What constitutes adequate elephant habitat, therefore, varies significantly.

Of the first five Kruger elephants translocated in 1979 to Pilanesberg, only one survived. Since initial translocation methods were limited, only young elephants could be transported, and therefore it is certain that all had been forcibly removed from their mothers and families well before normal weaning age. (As time has progressed, methods have been refined that permit larger and older elephants to be captured and moved.)

In 1992 Pilanesberg received two nineteen-year-old female former circus elephants brought back by their owner, Randall Moore.[15] Though orphans, said to have come from Kruger culls, as circus elephants Durga and Owalla had very different biographies from the young bulls'. At first, the orphans and circus elephants brought to Pilanesberg appeared to settle in despite the chaotic changes they had experienced. It was not long before the tiny park was home to more than eighty elephants. But suddenly, odd reports started to trickle in. Young male elephants had been seen harassing the two female matriarchs—behavior not only unbecoming of a young bull but highly irregular. And then there were the rhino deaths. Elephants soon were suspected, and video footage taken from a watchful tourist provided undeniable evidence: elephants were killing rhinoceroses.

In the ten years between 1991 and 2001, young male elephants in Hluhluwe-Imfolozi Park, located north of Durban, not far inland, reportedly killed fifty-eight black *(Diceros bicornis)* and white *(Ceratotherium simum)* rhinoceroses. Between 1992 and 1996, forty-nine white rhinoceroses in Pilanesberg National Park suffered a similar fate. Before these elephant-caused deaths, the average rhinoceros mortality was estimated at three per year.[16] Needless to say, the numbers of rhino fatalities were alarming to park officials, who feared that the entire populations might be lost. Adding to the gruesomeness of the gored bodies, the young bull elephants had been observed attempting to copulate with their victims. The data emerging from the two parks painted a disturbing, lurid picture. But there was more. As we saw earlier, Addo Elephant National Park has had elephant violence of its own. Unlike in other parks, there have been no recorded rhino killings by elephants, but Addo's male-on-male fatalities surpass all norms.

Addo Elephant National Park lies in the Sundays River valley of the eastern cape, adjacent to the blue waters of the Indian Ocean. Extending the park's domain into the ocean, its administration advertises that Addo encompasses five of South Africa's ecological biomes and is home to the Big Seven—the Big Five terrestrial game species plus the southern right whale and the great white shark. In addition, Addo has one of the densest populations of elephants, more than 450 individuals. Perched high on stilted viewing platforms, visitors can watch elephants sway past like boats on a sea of vegetation. The park is considered one of the most beautiful in South Africa.

Elephant demographics in all three parks—Addo, Pilanesberg, and Hluhluwe-Imfolozi—have been skewed. Much like what we saw in Uganda, each of the males at Hluhluwe-Imfolozi Park was less than fourteen years of age, and in Addo, older bulls were killed off: by 1940 all three founder males were dead. In 1931 the founder population consisted of two mature bulls and one immature male out of a total of eleven elephants. Between 1931 and 1938, there were eighteen births, and by 1954 fences were constructed around the park perimeter. This barrier virtually eliminated the elephant-farmer conflict that once had been the primary control on elephant numbers, and in the absence of human predation, elephant numbers increased to a relatively stable point.[17] There was a wrinkle, however.

The limited dimensions of the park created crowding and engendered a high level of physiological stress, which has been confirmed by measures of elevated corticosteroid metabolites. It is unsurprising that stressed, overcrowded male elephants are prone to male-on-male conflict. Interspecific aggression (farmers versus elephants) was traded for intraspecific aggression (elephant-elephant) that resulted in broken tusks and puncture wounds as well as the extraordinarily high level of male mortality.[18]

The behavior of the young bulls at Addo probably reflects cumulative effects of several factors that disrupt normal social life, including early trauma, abnormal attachment bonding, and conditions of chronic, elevated stress created by inadequate habitat. A key element, however, at all locales, is the absence of older bull socialization, resulting in a younger generation that has received little to no male mentoring.

Ethologically and psychologically, male elephants experience two critical phases that affect their brain and behavior. After leaving the natal family, when they enter into the very different society of male elephants, the young bulls learn how to navigate complex social interactions and how to regulate sensory-chemical systems of communication. They acquire the ability to signal appropriate behavior that helps maintain social order. This learning coincides with a second phase of major brain reorganization, one significantly slower than that which occurs in young females (a gender difference observed in many mammals).[19] Young male elephants are vulnerable to both learning and developmental disruptions whenever this second phase of all-male socialization is absent or truncated.

An attempt by a young male elephant to copulate with a rhinoceros is a behavior associated with extremely elevated states of central and autonomic arousal. This might have been a result of impaired sensory processing capabilities traceable to compromised development in the absence of older bull elephants, as well as early relational trauma from the cull. Impaired young bulls might not have been able to differentiate the sexual pheromone signals of a rhinoceros from aggressive pheromone signals of other male elephants. Neurobiologically, the bull's hyperaggression is consistent with an intense state of amygdala-hypothalamic sympathetic hyperarousal and a weakened higher right orbitofrontal inhibitory system associated with impaired developmental trauma.[20]

The abnormal fusion of affective, defensive, and fear-motivated rage with elements of a frustrated, intense sexual drive may have been exacerbated by the premature and sustained musth the adolescents entered in the absence of older bulls.[21] Although it is not certain that all rhinoceros deaths were coincident with elephant musth, periods of abnormal musth and peaks in rhinoceroses' mortality were indeed correlated.[22]

The young bull suspects were shot, but park rangers and researchers realized that a human-designed "reassembly" of elephant groups lacked the precision and orchestrated patterns of natural elephant society and took steps to offset this imbalance by bringing in six older bulls.[23] The essential developmental roles that mature bulls play was clearly demonstrated when their introduction to the park quelled the young males' abnormal musth cycles, premature sexual arousal, and interspecies aggression. Order was restored, at least temporarily, as the older bulls functioned much like corks in the emotional bottle, as external affect "regulators" in the absence of the young bulls' capacity to self-regulate.

It is difficult to pinpoint the exact cause of this adolescent aggression because the young males sustained multiple, major attachment disruptions and traumas. Any of the factors that they experienced—loss of mother and allomothers, premature weaning, witnessing family deaths, herd dissolution, translocation—could account for the deviation from normative behavior. But in the language of psychology and psychiatry, the bulls conform symptomatically to a diagnosis of posttraumatic stress disorder (PTSD).

The formal diagnostic criteria for PTSD are outlined in the *Diagnostic and Statistical Manual of Mental Disorders*-Text Revision (DSM-TR).[24] This text has become the "bible" of orthodox psychiatry and allied mental health professions. The American Psychiatric Association (APA), which publishes the DSM, is a professional organization whose members are drawn from diverse mental health professions. Individuals vary in how literally they take the DSM, but it functions as a lingua franca among mental health care practitioners and is referenced by nearly all of them.

It is a formidable volume, inches thick, and written and reviewed by scores of experts. Within its pages, the spectrum of human suffering is catalogued into discrete disorders used to systematize myriad symptoms into a plausible diagnosis. Some argue that its diagnostic criteria for PTSD are too restrictive, often excluding more complex versions and nuances, and that PTSD symptoms confusingly overlap with other psychiatric disorders such as depression and anxiety. Nonetheless, the DSM is suitable for our purpose of exploring psychiatric evaluation of elephant symptoms in human terms because it helps extend what we know about neuroethology into psychology. We do so, in much the same manner that the mirror test was executed, step by step, through clinical evaluation of each criterion comprising a diagnosis of PTSD.

The first criterion for PTSD refers to a traumatic event:

A. The person has been exposed to a traumatic event in which both of the following were present:
   1. The person experienced, witnessed, or was confronted with an event or events that involved actual or threatened death or serious injury, or a threat to the physical integrity of self or others.

2. The person's response involved intense fear, helplessness, or horror. Note: in children, this may be expressed instead by disorganized or agitated behavior.

By definition, what the young bulls experienced in the course of an elephant cull constitutes "actual or threatened death." Eyewitnesses describe infants screaming and running terrified by the noise of helicopters and guns and the bodies that are falling around them. Adie Read described to Chadwick an example of what the young elephant sees and hears:

> You see, at the last second before you open up, you get the pilot to do a low-level pass and drive them towards you before you sprint in. This gives you the initial surprise. When things go right the elephants mill. Total and utter confusion—they don't know what hit them. Just dust and shots and bodies falling down all over. We do most of the firing from just five to ten yards. Younger ones aren't keen on running away from the older ones. Our team left calves forty to forty-five inches at the shoulder [between about eight months and a year old] to sell to game ranches.[25]

The young bulls witnessed the deaths of their family members and feared for their own lives. Dame Daphne speaks of the profound terror that orphans show after seeing their mothers killed. Their experience is clearly filled with "fear, helplessness, and horror," and thus meet criteria A1 and A2.

The second DSM criterion for PTSD encompasses variations on "flashbacks" to a traumatic event:

B. The traumatic event is persistently reexperienced in one (or more) of the following ways:
   1. recurrent and intrusive distressing recollections of the event, including images, thoughts, or perceptions (**Note:** in young children, repetitive play may occur in which themes or aspects of the trauma are expressed);

2. recurrent distressing dreams of the event (**Note:** in children, there may be frightening dreams without recognizable content);
3. acting or feeling as if the traumatic event were recurring, including a sense of reliving the experience; illusion, hallucinations, and dissociative flashback episodes, including those that occur on awakening or when intoxicated (**Note:** in young children, trauma-specific reenactment may occur);
4. intense psychological distress at exposure to internal or external cues that symbolize or resemble an aspect of the traumatic event;
5. physiological reactivity on exposure to internal or external cues that symbolize or resemble an aspect of the traumatic event.

Obviously, these and other criteria for PTSD imply that traumatized patients can speak of their symptoms. In their own language, traumatized elephants do speak. Dame Daphne describes orphaned elephants who awaken terrified from nightmares and who exhibit wild-eyed panic and screaming in their early days at the Trust when they encounter humans and their machines. She comments on their fear when reminded of past associations of their trauma:

> If an orphan has had a bad experience during capture, that remains with them. Examples of this are Orok and Wasessa, both of whom had a blanket thrown over their eyes during the process of being captured, before being pounced on and subdued. Both these elephants have persistently refused to have anything to do with a blanket again. (Most of the Nursery elephants are blanketed during cold weather as a safeguard against pneumonia.) In Orok's case, we had to send him to Tsavo earlier than usual to avoid the cool season up-country, simply because he refused to be covered by a blanket. Wasessa, who is still in the Nursery, will have nothing to do with a blanket, and even tries to pull it off the others! Many orphans remember being transported in a vehicle to the airfield, and those that associate this with pain and trauma are very difficult to persuade into the trucks when it is time to

> transfer them to the Rehabilitation facilities in Tsavo. They have to be slightly sedated and then man-handled into the truck. Others that don't have such sinister connotations of a journey by road can be tempted in by their Keepers and their milk.[26]

At minimum, then, orphans who have had experiences similar to those of the young bulls conform to B2, B4, and B5, which satisfies the second diagnostic criterion.

The third PTSD criterion captures typical symptoms of avoidance and hypoarousal:

C. Persistent avoidance of stimuli associated with the trauma and numbing of general responsiveness (not present before the trauma), as indicated by three (or more) of the following:
   1. efforts to avoid thoughts, feelings, or conversations associated with the trauma;
   2. efforts to avoid activities, places, or people that arouse recollections of the trauma;
   3. inability to recall an important aspect of the event;
   4. markedly diminished interest or participation in significant activities;
   5. feeling of detachment or estrangement from others;
   6. restricted range of affect (for example, inability to have loving feelings);
   7. sense of foreshortened future (for example, no expectation of having a career, marriage, or a normal life span).

Because the pretrauma personalities and behaviors of individual elephants are generally not known, we must assume their symptoms are "not present before the trauma" based on observations of the behavior of other young elephants who have not suffered family disruptions and violence such as culls.[27] Elements C1–C3 are difficult to establish because the young elephants cannot report the details of their thoughts or feelings, nor whether they are having memory difficulties. However, in accordance with C4, C5, and C6, Dame Daphne has noted that many orphans are listless and apathetic, avoid social interaction, and uncharacteristically seek isolation; some nearly perish from grief. These orphans look and act detached and exhibit little of the engaged affection that is a hallmark of the species. Although we cannot determine whether elephants feel a "sense of foreshortened future," they do act depressed, with loss of appetite, apathy, inactivity, and lack of engagement with others—a significant departure from the social norms of elephant life and behavior analogous with that of humans who feel a lack of interest in the future.[28]

The final three DSM criteria speak to symptoms of hyperarousal, as well as the duration of all PTSD symptoms and their impact on daily life:

D. Persistent symptoms of increased arousal (not present before the trauma), as indicated by two or more of the following:
   1. difficulty falling or staying asleep;
   2. irritability or outbursts of anger;
   3. difficulty concentrating;
   4. hypervigilance;
   5. exaggerated startle response.

E. Duration of the disturbance (symptoms in Criteria B, C, and D) is more than one month.

F. The disturbance causes clinically significant distress or impairment in social, occupational, or other important areas of functioning.

Elephant families sleep together, and we must presume they sleep well under ordinary conditions. But Dame Daphne reports that orphans suffer insomnia, hence the presence of Keepers who sleep side by side with the infants every night. She writes,

> Post-trauma symptoms such as sleep problems, aggressiveness, depression, social and general apathy, listlessness, despair, nightmares and time out alone usually last longer than 1 month, depending on the personality of each individ-

> ual. These symptoms, once overcome and once the orphans learn to love their Keepers, are not repeated over the years. We can't truthfully say that we see a contrast between males and females, although I think that the little bulls are emotionally more resilient than the females who are born to be bonded into the family for life.

Dame Daphne also reports hypervigilance—"being on edge," outbursts, and nervousness—all symptoms that can last weeks.[29]

It is not possible to ascertain whether our young bulls exhibited an inability to concentrate or even hypervigilance and exaggerated startle response. In the case of the South African bulls, they did, however, exhibit general and specific hyperarousal, evidenced by their mounting and killing rhinoceroses—behavior that, as we have seen, is consistent neuroethologicaly with early developmental trauma and affective dysregulation.

Even with the more restrictive criteria of the DSM-TR, we can conclude—as clinicians might—that a diagnosis of PTSD is warranted based on what we know of the young bulls' history, neuroethology, and what has been observed in other young elephants with similar experiences. Obviously, there are variations with temperament and personality, but there is a convincing congruence between what models of stress and neuroethology predict and the behavior and other symptoms that are observed in the wild and in captivity.

For most elephant orphans not fortunate enough to be taken in by the Sheldrick Trust, there is no restorative social buffer. On top of the shock of seeing his family killed, enduring transport far from his natal herd, and missing the attention of a zealous constellation of aunties and siblings, a young bull has no father figures—no seasoned uncles to show him what being a big male pachyderm entails; no overnight camping trips that provide the training to hear and identify the sounds of the forest; no sideline tickets to watch how mature bulls know who is bluffing and who isn't, and, more important, who can afford to bluff and who can't. In short, in addition to a sudden, traumatic lapse in mothering, the young bulls suffered from fatherlessness, as did at least one of the three young Florida men who stood trial for the murder of Mrs. Diana Miller and the shooting of her husband, James.

Renaldo Devon McGirth, born on April 29, 1988, was found guilty of having committed murder "in a cold, calculated and premeditated manner without any pretense of moral or legal justification," and without being under the influence of any emotional or mental disturbance. Since Renaldo was eighteen years old at the time he was found guilty of murdering Diana Miller, he was tried as an adult and sentenced to death. The court found evidence that "clearly established that as early as at age 9 or 10, the defendant began a history of criminal activity leading him to have been found delinquent in juvenile court proceedings and being ordered to separate secure commitment facilities within the Department of Juvenile Justice."[30] The sentence and its public announcement would probably have been the extent to which Renaldo's case was known, if not for the extraordinary efforts of Lola González.

González, president of Accuracy Background Check (ABC), was the investigator and mitigation specialist who had been hired by the defense attorney, Candice Hawthorne, and appointed by the judge to provide background information dating from the time of Renaldo's birth to the time of the crime. González is held in high regard for her commitment, dedication, and compassion, as well as for her professional excellence. As a result, many difficult assignments come to her door, each having its own brand of tragedy. Renaldo's story is not unusual, nor was the troubling nature of his crime, but he stood out from other defendants whom González had encountered in her work.

As a mitigation specialist, González is responsible for helping the jury see and sympathize with factors over which the accused has had no control that have led or contributed to his crime. As González describes it, "The bottom line is that I must prove 'abandonment, abuse and neglect,' and somehow humanize a killer in the eyes of the jury so that he is not sentenced to death." The judge takes the jury's recommendation into consideration before the sentencing hearing.[31]

Proving abandonment, abuse, and neglect was not hard. A forensic psychological assessment performed before trial documented a wide range of symptoms in Renaldo's history: seizures, severe headaches, nightmares, nausea and vomiting, rapid heart rate, consistent mistrust of people, low body weight, restlessness, irritability, and feelings of hopelessness. Problems in childhood included night terrors, sleepwalking, repeated witnessing of violence, loneliness, severe punishment, an unstable home life, malnutrition, and habitual use of drugs.

While he was growing up, Renaldo never knew his father and was made the butt of jokes by his half-siblings (each of whom apparently had a different father). During his childhood, he rarely had enough to eat and, understandably, spent most of his time on the streets; child services workers reported multiple incidents of abuse and violence at his mother's home.[32] From supporting documents and conversations with Renaldo and his siblings, González learned that he had never questioned his mother's abusive behavior, nor had he mentioned it at trial, stating that he did not want his mother exposed to publicity. Though Renaldo claimed that "she was doing her best," it was reported that his mother at first refused to testify even when she was asked immediately before sentencing whether she wanted to address the judge in support of her son's plea for leniency.[33]

Throughout the mitigation interviews, when González questioned Renaldo about his father, the young man said that he had wanted to know since childhood know who his father was, but that his mother never told him.[34] When asked what had affected him growing up that made him feel the way he did, Renaldo wrote:

> Things that hurt me coming up as a child: Not being able to see my grandmother or talk to her when she passed away; I felt so alone, so now its like I don't wanna be alone; I was afraid of being alone; It was like I was mad at the world, I thought my father wasn't there because he didn't love me; . . . I always smiled, and I tried to hide behind an [*sic*] mask like I was okay but the truth was I was really hurting, then I felt short changed like on my birthdays there was never a party or

> even a celebration. . . . I remember when I would just lock myself in the room or the bathroom and cry but after a while I wouldn't do that I just shrugged everything off.[35]

Renaldo's plight touched González deeply. As the trial approached, González decided to launch, on her own, a focused search for Renaldo's father. After going through hundreds of documents in search of possible fathers' names and hitting as many dead ends, González finally asked Renaldo's mother, yet again, to name the father. Unbelievably, after so many refusals, the mother admitted that it was a man named Ron Grimes. She hurriedly sat down and started to systematically call, beginning with Florida residents, every Grimes in the phonebook who might be called Ron. No small feat. Discouraged, she came to the last name on the list. Crossing her fingers, she dialed the number and a forty-two-year-old Rondey Grimes picked up the phone.

At first, Grimes denied that he had any son named Renaldo. He did not even have any recollection of Renaldo's mother. It had been a long shot, but when González was finally able to jog Grimes's memory, the light went on, and Grimes started to put the pieces together. He told Lola that he had almost lost another son after the mother left, but, Grimes said, he was lucky and had been able to raise the boy on his own. The other son (who is nearly the same age as Renaldo) has lived a life opposite to Renaldo's. Grimes had provided a stable family—a new mother, Sharon, with whom he had a son and daughter, and a third boy, who had lost his parents and was adopted by the Grimeses. In contrast to Renaldo, his half-brother plays football and basketball, is coached by his father, does well in school, and attends church, all of which testifies to Grimes's dedicated and successful parenting. The parallel but diverging lives that the two boys have experienced leads one to wonder what different path Renaldo may have taken and who he might have become had he been able to come under the Grimeses' care as a growing boy.

Grimes immediately agreed to go to the jail to see whether the boy might be his son. He told González, "If that's my son, then he has a father." They met that night at the prison, and Lola knew she had found the right man. The family resemblance between the two foretold what DNA testing would eventually prove: Ron Grimes was Renaldo's father.

At their first meeting, González says, Grimes told Renaldo, "Son, if you did the crime, then you are gonna have to pay for it. And do it like a man." Grimes said that he had righted himself from a wrong turn taken in his youth and described such reform as possible and the only right course of action. From then on, the Grimeses, including Sharon and the four other children, attended Renaldo's trial and sentencing. Even Grimes's own father, Renaldo's grandfather, sat faithfully at the trial. Later, the Grimeses effectively adopted Quinton, a half-brother of Renaldo's (but not related to Grimes) who is also in prison.[36]

Grimes continues to mentor and stand by the young men, providing them what they never had: a father's love, attention, and guidance. Lola González has specialized her practice to help others like Renaldo who face death row. A nineteen-year-old currently awaiting trial and facing the death penalty, for example, was only a few months old when his father was deported back to Santo Domingo. The child was left with only his very young mother to raise him on her own. These young men have become González's cause. She has taken training offered by organizations such as the National Legal Aid and the Defender Association, and continues to support efforts for Renaldo and his brother because, she says, "They just didn't have a chance."[37]

Fatherlessness has been called the most "urgent social problem" of the century. David Blankenhorn has written extensively on the subject and notes that more than 50 percent of American children live in homes without a father. Not all agree with his conclusion that "fatherlessness is the most harmful demographic trend of this generation," but there is little dispute that it is a statistical reality reflecting a radical social change not limited to the United States.[38] By all accounts, the role of men, and even their presence in the lives of their children, has diminished with astonishing rapidity in communities across the globe.

Relative to the status of motherhood, an appreciation for the role of fathers and what young men face in the new social landscape has been slow in coming. In conventional terms, *mothering* elicits warmth and nurturance, whereas *fathering* tends to evoke in many human cultures contrasting qualities: strength, independence, and cooler, less emotional temperaments. Small wonder, then, that attachment studies generally have focused on the traditional mother-infant relationship, with fathers (or individuals fulfilling that role) only recently receiving much attention.

Boys are generally underexposed to men. The majority of teachers in the United States are female, and in the increasingly common single-parent families, that parent is typically the mother. Young men do not have the same access to familial older male role models that earlier generations enjoyed.[39] Writers and workers like Robert Bly, James Hillman, and Michael Meade have been instrumental in trying to offset the cultural imbalance and have sought to raise consciousness about the effects and challenges that a changed society has on the male psyche.[40]

In the United States, the new fatherless society was highlighted as the men's rights movement grew, with statistics showing strong correlations between fatherlessness and violent crime and poverty. The effects of human fatherlessness have been described in ways eerily reminiscent of the young elephant males. Steve Biddulph, author of the best-seller *Raising Boys,* writes, "Boys with absent fathers are statistically more likely to be violent, get hurt, get into trouble, do poorly in schools and be members of teenage gangs in adolescence."[41] Children growing up in single-parent households are more likely to have these issues. One Australian study found that 80 percent of rapists motivated by displaced anger came from fatherless homes.[42] Although correlation is not causality, most scholars of social psychology and policy agree that the near-epidemic destabilization of social processes has serious ramifications for mental and emotional well-being.

In 1999 approximately 80 percent of gang members in the United States were from fatherless homes. Indeed, one of the main factors contributing to the prevalence of gangs is the absence of older male mentors. "Most adolescent boys search for some form of masculine identity and male community wherever they can find it," writes Aaron Kipnis, a male psychologist and former president of the Fatherhood Coalition. The gang authority Steve Nawojczyk refers to gangs as the "5-H Club. They are homeless, helpless, hungry, hugless, and hopeless. . . . Gangs are the strongest where communities are the weakest."[43]

It is also true that in cases where crime is particularly violent, such as serial killings and torture killings, the perpetrators are statistically more likely to have been abused as children. James Gilligan, former director of the Center for the Study of Violence at the Harvard Medical School and director of mental health for the Massachusetts prison system, discusses the cycle of violence: "In the course of my work with the most violent men in maximum security settings, not a day goes by that I do not hear reports . . . of how these men were victimized during childhood. Physical violence, neglect, abandonment, rejection, sexual exploitation, and violation occurred on a scale so extreme, so bizarre, and so frequent, that one cannot fail to see that the men who occupy the extreme end of the continuum of violent behavior in adulthood occupied an equally extreme end of the continuum of violent childhood abuse earlier in life."[44]

A longitudinal study examining the relationships between adult ill-health and early childhood experiences showed that of almost twenty thousand subjects interviewed via questionnaire, 11 percent reported having been emotionally abused as a child; 30.1 percent reported physical abuse, 19.9 percent sexual abuse, and 12.5 percent witnessed their mothers being battered; 23.5 percent reported being exposed to family alcohol abuse and 4.9 to drug abuse; and 18.8 percent reported mental illness in the family: and the cycles repeat.[45]

In focusing on the children, we can forget to look back over their shoulders to the men and women who parent them. Their sorrows and pain lingered on and formed the seeds, indeed, many times constituted the direct cause, of their children's suffering. Before the twentieth century, South African men were forced to leave their families. Villages were bereft of men following European land seizures, the imposition of colonial political and economic structures such as taxes, and the labor demands of diamond and gold mines. Cultural breakdown undermined the individual who lost means and identity to function as before colonization. "The phrase *anginawo amandla* (I don't have power) expresses many forms of weakness among isiZulu speakers in KwaZulu-Natal. Sometime it denotes physical frailty, for instance if a person is tired or ill. More relevant is its use as an indicator of social weakness including—perhaps most pointedly—a man's inability to pay *ilobolo* (bridewealth)."[46]

These effects entail more than the single loss of a father: the elder entire male society and what it provided for the young crumbled. Mark Hunter, a researcher at the University of KwaZulu-Natal, argues that the concept of fathering and fatherhood should be seen in broader terms if these dramatic changes in family and society are to be understood: "The father clearly provided a role model for his sons, but the day-to-day socialization of young men appears to have taken place through peer groups or elder brothers and relatives as well as through the biological father. Certainly, fatherhood was a fluid category, with a boy's uncles referred to as *ubaba omkhulu* (bigger/elder father) or *ubaba omncane* (smaller/younger father)."[47] As a result of this exodus, today's young men "lack the father model with the tools to assume the authority and responsibility of being male in a patriarchal society."[48]

The themes of social breakdown, loss of homeland, and violence in South Africa mirror conditions in the United States and other communities, and they span time as much as geography. As one researcher wrote, "Our findings reveal the majority of South Africans do not experience just one traumatic event. Rather, individuals in South Africa experience multiple traumas. This finding highlights that traumas usually do not occur in isolation."[49]

Fathers, peers, uncles, older bull elephant groups, young bull cohorts, bull territories. Without the consistency of a loving and guiding family, there are few happy endings for humans or elephants. The young bull elephants at Pilanesberg were shot. Mrs. Miller is dead, her husband shattered from the loss. Renaldo is on death row. Quinton remains in prison. And so on and so on the story goes.

Today, after meeting his father, Renaldo studies and makes plans to get out of prison. He wants his story to be known so that others like him will do things differently. Renaldo wants to be able to start anew, this time with a family. But there is no guarantee. If his appeal succeeds, he may have a chance with Rondey Grimes's support and wisdom. He is not alone. There are glimmers of hope for rebuilding what has been dismantled. A number of local and national grassroots male mentoring programs have formed to provide what young men lack. Some at-risk youth-intervention programs such as No Guns, Unity One, and Barrios Unidos, staffed by "retired" gang members, bring experience and foresight to help young men forge a path other than violence.[50] Some young African elephant bulls are also getting a second chance. We will learn about how Dame Daphne repairs the wounded hearts and minds of young elephants and grooms them to take their place in the world as mothers, fathers, and leaders.

6

Elephant on the Couch

Case Study, E. M.

Presenting Complaints

E. M. (not her actual initials, case file identification 864858) is a thirty-two-year-old female (date of birth estimated 1976) who has been institutionalized since she was approximately two to three years of age. She came to the present facility in Dallas on December 14, 1986, and is being reassigned for a move to Mexico. Her current institution seeks to move her because of self-injurious behavior and persistent aggression in the form of outbreaks of attacks on staff. There are concerns that her depression following the loss of a companion will exacerbate these behaviors. According to staff, E.M has a history of "being very volatile behaviorally with low resistance to stress." She exhibits long-term physiological disor-

ders, including reproductive assays that reveal "atypical cycling, almost flat-line." She dislikes being confined to her room, where she exhibits a number of disturbed behaviors, including stereotypic rocking, banging and kicking her feet against the wall such that she loses her balance, spinning in circles, screaming, shaking and pressing her head against the ground, and swinging her leg back and forth repetitively. When she is very agitated, she exhibits what the staff calls her "wide-eyed" look.

E. M. has shown intentional aggression toward staff members and other residents, shoving them to the ground, pushing them against the wall, and lashing out to strike. Recently, while appearing calm, with her eyes half-shut, she hit one of the health care workers with no change in affect either before or after the event. Most outstanding is the "life-threatening" self-injurious behavior that has persisted for years, when E. M. strikes her right leg repeatedly. She also has severe, chronic injuries on her feet that have resulted in abscesses requiring intensive medical treatment. Associated with her self-injury is an unusual posture in which she places her leg on her elbow; the staff calls this her "funky lotus position." Phenobarbital and a long-term regimen of Acepromazine were prescribed with some success in attenuating symptoms.

History

E. M. was born in Africa and raised by her biological family with multiple siblings in a rural setting. Her cultural background indicates that she came from a matriarchal society where strong family values and collective interdependence are emphasized. At the age of two or three years, she witnessed her family massacred and was subsequently kidnapped. E. M. was then brought to the United States, where she was kept in confinement by a man who had a previous history of using welding torches, chemically treated metal prods on sensitive areas of the body, including the genitals, ropes, deprivation of water and food, beatings, and other forms of torture. It was common practice for him to hold a child under water as a form of discipline. Like others, E. M. was restrained with chains at all times except for short periods and given limited social contact.

At approximately age ten, she was transferred to the present facility in Dallas. Records state that E. M. was considered extremely unpredictable and that "behavior problems have become commonplace." Efforts to alter social contacts and living quarters were considered to be ineffective. Androstenediol levels were assessed at three times normal. She showed extreme aggression to another, younger (eight years old) female resident, such as running after her and shoving her into the wall.

Self-injury and stereotypy developed in 1995, during a period when extensive construction work was being conducted at the facility. In 1998 she was no longer physically restrained but was kept within a barred enclosure unless sedated for medical treatments. E. M. continued to be aggressive with staff and other residents. In 2000, when the other, younger female was transferred, E. M. became "moderately depressed," "rocked constantly," and became anhedonic. Concern for long-term side effects of Acepromazine prompted a gradual "weaning" off the medication, but rocking and other stereotypic behaviors subsequently increased. Periodic Acepromazine treatment has been resumed to date for recurring symptoms.

For example, from April 30, 2007, to June 12, 2007, E. M. was put back on Acepromazine in anticipation of resumed construction at the facility and from April 30, 2008, to May 25, 2008, because of the illness of another female resident of approximately the same age, "K.," who had been transferred in to the facility. K. came from a similar background and was considered to be "confident" and able to assert herself. E. M. was submitted to physical restraints.

On September 19, 2007, E. M. ingested a substantial amount of a rubber toy. She exhibited other indications of eating disorder. After a day of diarrhea, she was, according to staff records, weak, lay down often, and had difficulty rising and walking. At time she lay on ground incontinent, kicking her feet; at other times she leaned against the wall in obvious pain. Finally the staff gave her ketoprofen and butorphanol. The following day a plastic bag was found in her feces. When K. became ill and finally died, E. M. showed extreme distress, increased instability, and disturbed behavior, including head banging, screaming, running through her room, and anorexia. She insisted on staying near her companion as much as possible. Acepromazine was administered, but her legs buckled, she was unable to stand, and she sat in one place motionless for four hours. Over the next days, she exhibited stereotypic rocking. A decision regarding E. M.'s possible transfer is pending.

A painful, but not atypical, excerpt from the medical logs of a mental institution, of the kind thankfully less common today than in the past, when mental illness was stigmatized far more than now, but not atypical in the case of elephants. It may be no surprise to learn that "E. M." is a middle-aged African elephant whose real name is Jenny. Her case history was compiled from public records of the Dallas Zoo, where she resides. A discerning eye might have noticed the acepromazine prescription; "ace" is a psychotropic drug largely discontinued for humans after the 1950s because of its profound side effects, but not uncommonly used to sedate animals. Other than that, Jenny's description and residence might very well have been the inspiration for Ken Kesey's acclaimed novel *One Flew over the Cuckoo's Nest.* Elephant captive housing may not be referred to as a pachyderm asylum, but from the perspective of elephant sensitivities and sensibilities, it bears more than a superficial resemblance to old-style human institutions.

In comparison with most other confined elephants, Jenny's records are quite lengthy. There is great detail and a series of letters from multiple veterinarians called in to evaluate her. Many of Jenny's symptoms are fairly typical in zoos; stereotypy (repetitive swaying, incessant chewing, or other behavior to cope with the distress of confinement), arthritis, hyperaggression, depression, foot disease, and overall progressive degeneration of mental and physical health are more often the rule, not the exception, but Jenny's case received particular attention because of her potentially life-threatening self-mutilation.[1] Self-injurious behavior is not unknown. An elephant veterinarian of the Oregon Zoo in Portland, Michael Schmidt, observed that a female elephant continually pulled on her teat until it gradually lengthened and distorted in shape, and a bull repeatedly broke off his tusks under the continual stress of confinement.

Despite the affectless, understated tones in the medical letters, the sheer volume of expert interest and documentation illustrate the level of concern. Nonetheless, as with others in captivity, a lot is missing from Jenny's history, symptoms, treatment, and handling, either because they were not noticed, never recorded, or considered best left undocumented. Elephant welfare is a recent concern compared with the animals' history of captivity in the United States: elephants had been imported to the United States for almost two hundred years before welfare and protective regulations were established.

The first elephant known to have graced U.S. soil was imported from Bengal at less than three years of age in 1796 by Jacob Crowningshield.[2] Soon after, elephants became a form of institutionalized entertainment when the farmer Hachaliah Bailey imported a young African elephant. A harbinger of tragedies to come, Old Bet did not reach the age her name implied; she was shot to death in 1816, as was her successor, Little Bet. Nonetheless, traveling menageries became wildly popular. The famed circus founder P. T. Barnum launched an expedition in 1850 that tracked down ten elephants in the jungles of Ceylon (now Sri Lanka). In 1882 Barnum purchased a huge African elephant named Jumbo from the London Zoo and brought in hundreds of spectators, for whom the elephant became the quintessential symbol of circuses. Jumbo was killed by a freight train in 1885. After Barnum's own death in 1891, the circus menagerie was purchased by the rival Ringling Brothers show to form the Ringling Brothers and Barnum and Bailey Circus, which today is owned by Feld Entertainment.

By the twentieth century, elephants had created a sort of proto-pop-culture sensation, long before Barbie dolls or *Star Wars*. Despite the logistics of capture and transport of a several-ton individual across thousands of miles and varying climates, by 1952 there were 264 elephants in the United States—124 in circuses and the rest in zoos, which had joined in the business of elephant exhibition. All but six of these elephants were females, because controlling the power of a full-grown bull, particularly during musth, was an even bigger challenge.[3] Today, out of more than 14,000 Asian elephants worldwide, one-third are held in captivity. Facilities in North America maintain approximately 600 Asian and African elephants. These numbers do not remain static. A substantial mortality rate, together with tightening restrictions on procurement of wild-caught elephants, limits the U.S. elephant population. Captive breeding programs have been developed and refined to offset the difficulty in maintaining and growing a captive population.

The Animal Welfare Act (AWA, 1970) and the Endangered Species Act (ESA, 1973) are the two main pieces of federal legislation governing elephant welfare and the safety of humans working with elephants in the United States. From an elephant's point of view, the laws offer little protection. Tellingly, the AWA is administered by the U.S. Department of Agriculture (USDA) and administered through its Animal and Plant Health Inspection Service (APHIS). There are no specific guidelines tailored to care of elephants, with the exception of tuberculosis testing, treatment, and autopsies.

The number of inspectors assigned to monitoring and enforcing the AWA is miserably inadequate. In 2003, one hundred or so inspectors were responsible for approximately ten thousand to twelve thousand facilities nationwide, an increase from 1999, when only sixty such positions were funded.[4] Most inspectors have no special expertise with elephants. Records concerning elephant health and transfers are not readily open to the public, which means that animal protection groups cannot make up for deficits created by government understaffing. What confounds effective tracking is that even though the law requires any exhibitor to keep written records of animal acquisition and veterinary care, public access via the Freedom of Information Act (FOIA) is often unavailable. USDA inspectors review case files on site, but the agency does not normally retain copies, which means that those records are not available through FOIA. Elephant studbooks (records that log statistics per individual, such as known or estimated birth date, birthplace, and ownership) provide the bare minimum of information, and their accuracy is often suspect. Confusion arises because of identical names for numerous individuals (Jenny is a common name, for example) and changes in exhibitor's names. As a result, many elephants are bought and sold without proper recordkeeping, and the fact that there is no single, centralized repository of data translates to a shortfall in elephant welfare. All of which conspires to undermine elephant protection. Then there is the matter of having elephants in captivity at all.

Once a species is listed under the Endangered Species Act, any person subject to the jurisdiction of the United States is disallowed from "taking" that species. *Taking* covers a range of actions including harassment, injury, and killing. However, all elephants are not created equal. Asian elephants are considered an endangered species, but African elephants are classified as "threatened," which gives them much less protection.

There is another dimension to the regulations: human safety. When there are occasions for a "dangerous animal" to come into contact with or be near the public during elephant rides or performances, the presence of an experienced attendant is required. This person is responsible for care, housing, handling, sanitation, nutrition, water, and providing adequate rest between performances and protection from extreme weather and temperatures. Records from the 1990s document more than one hundred deaths or injuries to elephant personnel, audience members, or passersby. Some keepers have been cited for violations including inadequate care for tuberculosis and other illnesses.

Obtaining permission to own and show an elephant is fairly simple. Permission is granted with a license that costs somewhere between $40 and $310, depending on how big the circus or display is, how many animals will be displayed, and what kind of business the enterprise is. Regulated animal exhibits include zoos administered by governments, private foundations and individuals, corporate businesses, as well as diverse performance acts including the familiar circus and mom-and-pop roadside menageries. None of the existing regulations take into account critical details concerning elephant biology, ethology, and ecological requirements, which is why zoos and other institutional captivity produce elephants with such dire mental and physical ailments. Jenny's first owner, Roman Schmitt, who kept and trained her for ten years before selling her to the Dallas Zoo, was required to meet these standards; Jenny's condition is testimony to the lapses in government regulation appropriate to elephant needs.

In August 2004 an eight-month-old infant elephant, Riccardo, fell off a nineteen-inch round pedestal during a circus act and broke both hind legs—an injury so severe that he was euthanized. Another young elephant, four-year-old Benjamin, had a fatal heart attack in a pond while trying to evade a trainer armed with a bullhook, who had beaten him repeatedly. A government inspector determined that the bullhook had precipitated Benjamin's physical harm and ultimate death.[5]

A variety of tools, including food and water deprivation and other methods of causing distress, pain, and injury, are used to force an elephant to live in captivity and submit to total control by the trainer. This process is called breaking. Circuses and some zoos employ a "dominance-based free contact" system grounded in principles of physical and psychological coercion. Ankuses, while used less in zoos that have adopted so-called protected contact—in which elephant and handler do not occupy the same space, and positive reinforcement is used to encourage behavior—have been a standard.[6] Veterinarians observed cumulative and numerous ankus wounds on Pet, from the Oregon Zoo. As late as 2005 a veterinary exam showed that she had a wound under her trunk that "corresponds with probable ankus tip poke wound (prompting trunk lifting) and keepers believe ankus injury is the cause."[7]

Typically, the breaking process begins by forceful removal of infants from their family units, followed by bodily immobilization, beating, and starvation and other deprivation until the elephant accepts the trainer as his or her "master." A broken elephant is one who has ceased active resistance against restraint and confinement. Negative reinforcement techniques are a part of regular training: bullhook beatings for poor performance, displays of resistance, and/or unapproved socialization with other elephants. The severity of negative conditioning through the breaking process allows the trainer later to use relatively little force in performances. The elephant trainer and circus consultant Alan Roocroft, who has worked at multiple zoos and teaches elephant management, writes: "When corporal punishment is administered to an elephant, it has to be fairly forceful in order that it is perceived by the elephant to be punishment at all. . . . The trainer must now intimidate the animal in order to acquire a dominant position. . . . Restraining a potentially hostile elephant needs at least a crew of eight, preferably in order to insure sufficient 'muscle' is available. Once immobilized, the elephant may be the object of punishment in the form of blows with a wooden rod."[8]

A former keeper describes a typical disciplinary act:

> The elephants were ordered to hold their trunks up in the air in "salute" position while handlers were performing daily care routines. For the elephant Tina . . . the routine took longer than usual and her trunk began to cramp. Finally, Tina could no longer hold it up. When she lowered her trunk, a stream of accumulated mucus leaked out, indicating that Tina was unable to breathe through her trunk. Nevertheless, a supervisor witnessed this breach of command (Tina lowering her trunk), and ordered Tina to be chained for discipline. The keeper was told to "work her over," which meant running her through her commands—forcing her to move from side to side while on chains—and striking her with the ankus when she did not comply quickly enough. The keeper was not being forceful enough with Tina for the supervisor, who took over the session.[9]

An elephant veterinarian for more than thirty years, Schmidt maintains that "the modern zoo is as dangerous for elephants as it always has been." The use of the term *dangerous* is interesting. Of late much has been made about the necessity of captive breeding of elephants because their conditions in the wild are so threatening. Indeed they are; however, statistics comparing free-ranging elephants with those in captivity demonstrate that confinement is not only dangerous but lethal. One of the National Zoo's leading researchers on elephant reproduction, Janine Brown, predicts that the severity of problems with zoo elephants—more than 30 percent infertility; high infant mortality, including infanticide; neurotic and stereotypic behavior that includes calf rejection, self-mutilation, and intraspecific aggression leading to deaths; serious foot and weight problems—will result in "the extinction of all elephant species in North American zoos within only a few decades."[10]

In a study of forty-five hundred elephants in captivity—approximately half of the global captive population between 1960 and 2005—researchers found that the median lifespan of African elephants in captivity was 16.9 years, compared with 56.0 years in free-ranging Kenyan elephants. For Asian elephants, those in zoos showed a median lifespan of 18.9 years, those in captivity in Burmese logging camps 41.7 years.[11] Researchers suggest that elevated adult mortality of Asian elephants in captivity occurs because of stressful conditions during gestation or early infancy and that mortality rates for elephants that are moved increase by 50 percent for four years. Furthermore, "survivorship tended to be poorer in Asian calves removed from mothers at young ages," and "within zoos, captive-born Asians have poorer adult survivorship than wild-born Asians. . . . This is a true birth origin effect: Whereas zoo-born elephants are more likely to have been born recently and to primiparous dams [individuals giving birth to their first infant], neither dam parity nor recency predict adult survivorship. . . . Because the median importation age of wild-born females was about 3.4 years, this suggests that zoo-born Asians' elevated adult mortality risks are conferred during gestation or early infancy." Overall, the researchers conclude, interzoo transfer, maternal loss, and health and reproductive problems in zoo elephants cause obesity and stress. In short, "bringing elephants into zoos profoundly impairs their viability."[12]

Foot ailments are among the most common and debilitating symptoms of elephants in zoos and circuses. In contrast to the almost continual movement of free-ranging elephants on grass and soil substrates, some elephants in captivity spend twenty hours a day on unyielding concrete and asphalt, with very little room and exercise. A thirty-plus-year study by the Amboseli Elephant Research Project (AERP), based on more than thirty-four thousand sightings of wild elephant groups containing up to 550 individuals, found no chronic foot or weight problems in the Amboseli elephant population. In contrast, a survey derived from public records by the animal protection organization In Defense of Animals (IDA) showed that in forty-six AZA accredited zoos, holding 135 elephants, 62 percent of the elephants have severe foot disease and 42 percent have joint disorders. Toni, a wild-caught Asian elephant at the National Zoo, was euthanized at the premature age of thirty-nine. Her feet, like those of most other zoo elephants, were cracked, infected, and had nail abscesses; she had lost more than nine hundred pounds as a result of these and other cumulative illnesses, leaving her emaciated and debilitated. As the veterinarian Mel Richardson, who observed Toni shortly before her death, noted:

> Whenever possible we as veterinarians are trained to prevent pain and suffering, not just treat it. Why are the veterinarians at the National Zoo not preventing the painful degenerative arthritis in their elephants like Toni and Ambika? They cannot! Because the cause of the crippling degenerative joint disease is the exhibit itself: the concrete; the packed unyielding abrasive substrate inside and outside; the lack of exercise and normal use of the elephant's feet and limbs—climbing, digging, walking, wading into streams, kicking logs, and foraging. . . . [Elephants] evolved to travel miles each day on uneven natural substrate using their feet to find and apprehend food. To keep them healthy we must provide that opportunity as well. The zoo exhibit itself is the cause of the Degenerative Joint Disease. The zoo exhibit itself is killing her.[13]

The pressure to produce more elephants in captivity often overrides medical wisdom and humane care. Pet, the elephant at the Oregon Zoo, was euthanized at fifty-one. Her feet were so damaged that she was forced to wear sandals and used her trunk as a crutch. Having lived decades on concrete surfaces, she developed severe degenerative joint disease in all four legs. X-rays showed that she had complete "collapse of intercarpal joint spaces, bone lysis and osteophyte formation": her feet were effectively breaking down. Nonetheless, despite her fraility, Pet was given a "transrectal ultrasound/reproductive exam" on May 4, 2006, only three months before she was euthanized. She was found to have "no reproductive abnormalities" but would "not be bred due to age and severe DJD."[14]

Birth and family, so pivotal in elephant culture, can bring heartache to those in captivity. The experiences of Lisa, a female African elephant at the Oakland, California, Zoo, considered one of the United States' most progressive, illustrate some of the issues surrounding captive breeding and birthing.

Lisa became pregnant three times with the resident bull, Smokey. Sadly, all three pregnancies turned tragic. Video coverage shows Lisa's labor and the birthing process is consistent with AZA protocols: "When the first signs of labour appear, the elephant handlers will tether the elephant on 3 or 4 leg restraints (chains or ropes). . . . The newborn calf will be immediately removed from the mother. . . . The rest of the herd will be in adjacent stalls to avoid interfering with the elephant care staff but still observe the birth and newborn calf. . . . They may also need to be tethered if they become agitated. . . . Extreme caution must be utilized at all times. . . . The mother may not respond to commands as well as she usually does."[15]

The video shows Lisa in labor with her legs chained. The moment the baby, a male later named Dohani, emerges and slides to the cement floor covered in bloodied afterbirth, a zoo staff member pulls him away with a long metal hook. Over the next several hours, Dohani is subjected to a battery of exams and procedures—vacuums to suck out material from his trunk, syringes to extract blood, and instruments to weigh him. As time passes, he clearly becomes more stressed, and his eyes begin to bulge from the strain and strange noises and activities as staff busy themselves with the various medial tests. When Lisa is on camera, she is pulling desperately against her chains to reach her baby, vocalizing, extending her trunk through the barrier. At one point she breaks through one set of chains. Eventually mother and child are united, but their blissful time is wrenchingly short. Days later Lisa punctured Dohani fatally with her tusk.

Infanticide is so common in zoos that personnel almost routinely remove a baby from the mother immediately after birth. In 2008 a mother elephant in Berlin appeared to try to drown her infant before zoo personnel intervened. Extensive measures are taken to preempt infanticide. Often it is not clear whether a mother elephant injures or kills her infant intentionally or by mistake. Why Dohani was gored by his mother is unknown, but the outcome is consistent with the 2008 survey that documented the impacts of zoo captivity on elephant mortality.[16] Dohani's death was yet another tragedy for the young mother. A few years before, Lisa's infant Kijana had died from a herpes virus at less than one year old, and another baby was stillborn, possibly because of salmonella. Smokey died of unknown causes.

In contrast, out of fifteen hundred births observed in the wild by the Amboseli research group, no cases of infanticide or calf rejection have been reported. Similarly, there have only been two cases of infertility out of 558 females more than ten years of age. The IDA zoo survey showed that 73 percent of captive pregnancies involved complications and reproductive disorders, with nine stillbirths and, as with Jenny, an extremely high rate of early infertility, including elephants who have "flat-lined," become infertile as early as age twenty-three to twenty-eight years, with sixty stillbirths.

All of these behind-the-scenes data paint a very different picture from that portrayed in circus flyers, Hollywood movies, and ads for exotic elephant rides. The brightly arrayed performing elephant with the delicate ballerina circus rider perched atop transforms into a grotesque caricature. The outburst of Tyke, an elephant who in the 1990s at a Honolulu circus performance suddenly charged and deliberately, repeatedly crushed her trainer in front of horrified circus crowds, horrifies us now for a different reason. Jenny, Tyke, and others begin to resemble the prisoners of whom psychiatrist James Gilligan writes: "These men are 'dead' inside. For how else could they possibly murder others and mutilate themselves as they do—unless they had no feelings?"[17]

Tyke is dead (she ran out of the circus tent and was shot with nearly one hundred bullets in the Honolulu streets), but, unlike Gilligan's prisoners, she was far from dead while she struck out at her trainer. Tyke's rage and Jenny's flesh gouging speak of vital pain. The germ of self that existed before the birth of unbearable pain is buried deep within a lacerated mind.

We now have enough background to understand underlying causes for Jenny's condition and symptoms and make a diagnosis. Jenny's case history was circulated to five mental health professionals without foreknowledge of her species; she was identified only as a thirty-two-year-old female with the initials E. M. Each clinician was an expert in his or her field and seasoned in working with people with a range of symptoms and issues. The resultant diagnoses were remarkably similar. All five diagnosed "E. M." with PTSD, and their suggested treatment plans converge. A therapist with a thirty-five-year practice suggested that Jenny had "severe PTSD on Axis 1 and probably Antisocial Personality on Axis 2. The Axis 2 dx [diagnosis] would likely be secondary to the PTSD as the primary precipitating factor. Treatment plan would likely involve significant use of medications and a very carefully and tightly controlled behavioral approach within a very intimate relational context, i.e., a single therapist, who adapts to EM first, then attempts to help EM adapt reciprocally."

A European psychiatrist specializing in schizophrenia concurred, but consistent with a psychiatric versus purely therapeutic perspective, emphasized psychopharmaceuticals: "PTSD with comorbid pica. I suggest putting her on SSRI [selective serotonin reuptake inhibitor], e.g. Sertraline 50 mg. Could be better than antipsychotics, especially more helpful in ameliorating the stereotypies." The other three also diagnosed PTSD with some variant, including a diagnosis of developmental trauma disorder (DTD) by a pediatric specialist. DTD is a more defined replacement of reactive attachment disorder (RAD). RAD has been used to describe children who have experienced severe abuse, neglect, or significantly impaired attachment patterns.

All of these diagnoses derive from the *Daignostic and Statistical Manual of Mental Disorders* (DSM). They may look the same, but the jumble of acronyms and terminology can be confusing for those unfamiliar with the maze of clinical verbiage. Before going on, let us explore some of the less than transparent terms.

Substantial controversy surrounds the DSM. For one, symptoms can be related to multiple causes and vice versa, and not every person's behavior and mental state can be neatly labeled and put into a discrete box or category. Depression is common to a number of underlying causes, as are hostility, violent outbursts, asocial behavior, addiction, and dissociation. Furthermore, many claim that ascribing pathology to symptoms amounts to blaming the victim and imposing value judgments reflective of a certain social norm. For example, as late as the 1970s homosexuality was included in the DSM as a disorder. As a result, some practitioners largely ignore the DSM, using it only to satisfy insurance forms requiring a codified, numbered diagnosis. Nonetheless, the DSM remains the cornerstone reference, which explains why the five clinicians consulted about Jenny's conditions, each from a different subfield of psychology, all used the same terminology. The differences in their diagnoses, though, are important to the question of whether the use of human psychiatric principles is suitable for an elephant.

In their diagnoses, the therapists named PTSD as an "Axis I" diagnosis. The DSM diagnoses are organized according to a "multiaxial" system. Because a person can have several issues at once, this system was created to help sort through the web of symptoms. Axis I is considered the focal and most obvious condition affecting the individual, while successive axes are secondary or pertain to other potentially contributing factors, such as medical conditions or environmental issues. For example, job loss, divorce, and other events affect and are affected by a person's mental state and behavior.

Across the board, Jenny's Axis I diagnosis was determined to be PTSD, specifically Complex PTSD. Complex PTSD and DTD are often diagnoses for individuals who are traumatized early in life, coupled with sustained psychological insult. The majority and primary etiology of Jenny's symptoms were ascribed to, in the words of one therapist, "multiple, abusive, torturous, traumatic relationships." Similar to the South African young bulls, Jenny suffered from early maternal and endured "sequential losses." Unlike theirs, however, Jenny's series of traumas was compounded by moving between "multiple residences, never having a home," leading to what the therapist viewed as "lack of trust and grieving that underlie much of her behavior." Three of the clinicians who reviewed Jenny's case files included additional, differential diagnoses.

While PTSD is considered primary, one therapist categorized Jenny's aggression to staff and her conspecific residents as an Axis 2 (secondary) "antisocial personality." Antisocial personality is described in the DSM as behavior "that is not due to a mental disorder. . . . Examples include the behavior of some professional thieves, racketeers, or dealers in illegal substances." The reason given by the therapist for including antisocial behavior as part of Jenny's diagnosis is that "Axis 2 personality disorders are more commonly thought of as originating in the psychodynamics of early childhood. So E. M.'s Axis 2 diagnosis—her hostility and aggression—which likely arises from the PTSD experiences is, therefore, 'unorthodox' because the Axes are related." This is where the art of a clinician becomes critical and the smoky ambiguity of psychological diagnosis starts to surface.

One psychiatrist thought focal attention should be directed at Jenny's "pica" behavior, the ingestion of nonnutritious foods, in her case the rubber tire, for more than one month.[18] Pica is a syndrome not infrequently observed in zoo animals and is considered a disorder originating from behavioral or biochemical sources. There are various hypotheses as to its source, including poor nutrition, but its prevalence in captivity and effective treatments indicate that much of the occurrence relates to stress and trauma. In humans, pica is found in individuals who have suffered abuse as children.

When the second therapist was questioned as to the psychiatrist's diagnoses, he replied: "The self injury and pica invite borderline personality into consideration, and certainly there are psychotic features which invite schizotypal personality disorder as well. The dissociative disorders' symptomatology in the DSM doesn't match up all that well in E. M.'s case, even though dissociation is a feature of PTSD. The anti-social features seemed most prominent, but given E. M.'s traumatic history, any and all of these diagnoses are appropriate." In other words, the circumstances of her traumatic past and present are so extreme that they can create a number of different constellations of symptoms and hence diagnoses secondary to PTSD.

A fourth clinician, whose practice has specialized in Holocaust survivors, chose to use not the hierarchical format of multiaxis diagnoses but rather a "nonaxial" format which lists possible diagnoses, all of which focus on trauma: "An anxiety disorder such as . . . Posttraumatic Stress Disorder based on her childhood traumas . . . witnessing death of family and low resistance to stress. Dissociative Disorder Not otherwise Specified . . . for being subject to prolonged and intense coercive persuasion . . . kept and tortured." The clinician also specified a mood disorder such as recurrent major depressive disorder with catatonic features—maybe for her "funky lotus position, which may be motor immobility, bizarre posture, resistance, can include self harm and that of others." A fifth therapist brought attention to the possibility of exacerbating effects from the sustained medication regime Jenny had been prescribed. "Medicated to control behavioral and emotional issues. Mood seems to be primarily vacillating between anger with aggressive outbursts to general withdrawal from life."

A maze of terms and concepts indeed. Yet overall, the consensus points to a series of traumas as the agents of Jenny's behavior and mental states. In fact, when we examine the list of possible PTSD symptoms in the DSM, far less seems to be excluded than included. Associated disorders and symptoms include self-destructive behavior, dissociative symptoms, somatic complaints, despair, hopelessness, feelings of permanent damage, a loss of previously sustained beliefs, hostility, social withdrawal, constant feelings of threat, impaired relationships with others, a change from the individual's previous personality. PTSD is also associated with other disorders: major depressive disorder, panic disorder, obsessive compulsive disorder, and generalized anxiety disorder, among others. All these technical discussions make for a complex diagnosis. But the second psychologist probably put it most succinctly: "Maybe the most appropriate dx [diagnosis] of all is 'broken' or 'shattered.'"

What insights, if any, have we gained from this journey through the technical warrens of psychiatry? All this seems like a lot of trouble to state the obvious. Elephants don't do well when they are made captive and sustain the deprivations imposed by zoos and circuses—something that psychologists like Harry Harlow and others established prima facie years before with monkeys and chimpanzees. But an important step has been made. Beyond neuroethology that predicts that an elephant who undergoes traumas from childhood through adulthood will exhibit behavior comparable to humans who have undergone similar experiences, we learn that psychiatric diagnoses fit across species. In so doing, science has travelled from trans-species neuroethology to trans-species psychology and psychiatry. The fact that PTSD has been diagnosed is not the only news; the condition is considered to have an evolutionary basis shared across species.[19] What is more newsworthy is that the clinical methods, tools, and criteria developed for humans can be translated to elephants. But while the foundation is laid for legitimizing a common psychobiology for elephants and humans, there is much more to explore about the elephant psyche, and thousands of animals continue to live out their lives behind walls or bars.

The new executive director of the Columbus Zoo and Aquarium, Jeff Swanagan, spoke with candor about the new reality: "Everybody thinks the zoo is all about animals but it's not. This is primarily about the people." An ever-lengthening list of cities and communities have sought to ban exotic animal acts, and more zoos are closing their elephant exhibits, including the Gladys Porter Zoo (Brownsville, Texas), the Detroit Zoo, the San Francisco Zoo, Chehaw Wild Animal Park (Albany, Georgia), the Henry Vilas Zoo (Madison, Wisconsin), the Louisiana Purchase Gardens and Zoo (Monroe, Louisiana), Mesker Park Zoo (Evansville, Indiana), the Frank Buck Zoo (Gainesville, Texas), the Sacramento Zoo, the Lincoln Park Zoo (Chicago), and the Alaska Zoo (Anchorage). Those expected to close their elephant exhibits soon include the Philadelphia Zoo, Lion Country Safari (Loxahatchee, Florida), the Bronx Zoo, the Santa Barbara Zoo, and Buttonwood Park Zoo (New Bedford, Massachusetts). Dulary, a resident of the Philadelphia Zoo, was moved to the Elephant Sanctuary in Tennessee, and Maggie, a resident of the Alaska Zoo, was moved to PAWS, but unfortunately, in most cases, elephants are relocated to other entertainment facilities or zoos. At the same time, pressures to expand captive breeding programs have increased. The Center for Elephant Conservation was established in 1995 by the Ringling Brothers and Barnum and Bailey Circus in Polk City, Florida, as a private breeding facility designed to increase captive populations. Other breeding facilities include the Indianapolis Zoo, the Houston Zoo, the St. Louis Zoo, Dickerson Park Zoo, Rosamond Gifford Zoo, the Oregon Zoo, Disney's Animal Kingdom, and the San Diego Wild Animal Park.

Circuses face even more resistance to traditional exhibition of elephants. The entertainment industry has enjoyed a loyal following for decades, but overall attendance numbers have decreased significantly, and a number of states and cities have implemented restrictions, some localities even banning animal acts or legislating against such practices as the use of bullhooks. Both the Ringling Brothers and the Clyde Beatty circuses nearly went out of business in the 1960s, when Feld Entertainment revived its market by taking its circus to indoor arenas. Many circuses have dropped animal shows from their programs, instead staging elaborately choreographed shows on the model of Cirque de Soleil. But Ringling, Clyde Beatty, UniverSoul, and Carson and Barnes continue to use elephants, although this too is being challenged in federal court with charges of abuse against Feld Corporation.[20]

Captivity as a socially viable institution is no longer a given. Deep questioning of the ethics of captivity has taken root, which is good news for the residents of zoos and circuses. Nonetheless, for many in the United States and throughout the world, the damage has been done: for the foreseeable future, a steadily growing number of elephants will need physical and psychological rehabilitation and sanctuary—and the same will be true for hundreds of other species. Who will take them in? And can therapies be designed and implemented that can help these individuals recover?

For Jenny, unfortunately, the opportunity may never come. The Dallas Zoo has decided to keep Jenny after all, and will house her with five or more elephants in a five-to-six-acre "African Savannah" exhibit, where she will wear a global positioning system (GPS) unit to record her movements.[21]

1. Happy passes the mirror self-recognition test at the Bronx Zoo. Photo courtesy of Diana Reiss, Joshua Plotnik, and Frans de Waal

2. Pat Derby, founder of Performing Animal Welfare Society (PAWS), and one of her beloved elephants, called 71, who has since died. Photo courtesy of Pat Derby, Performing Animal Welfare Society (PAWS)

3. Dame Daphne Sheldrick, founder of the David Sheldrick Wildlife Trust. Photo courtesy of David Sheldrick Wildlife Trust

4. A free-ranging, intact elephant family in Amboseli National Park, with a social life that is rich and complex. Photo courtesy of Cynthia Moss

5. An elephant baby is raised by a constellation of allomothers. Photo courtesy of Cyril Christo and Marie Wilkinson

6. Billy Jo was raised as a human child until his teen years, then was sent for use in biomedical experiments. His compromised development as a cross-fostered chimpanzee contributed to lifelong difficulties in trauma recovery. Photo courtesy of Gloria Grow, Fauna Foundation

7. The brutal training of circus elephants, such as these in the Ringling Brothers and Barnum and Bailey Circus annual elephant walk in Washington, D.C., has been challenged by a lawsuit in the U.S. District Court. At issue is whether the 1973 Endangered Species Act legally permits the circus to chain its elephants and control them with bullhooks. Photo courtesy Amy Mayers

8. By definition, orphaned elephants all have undergone trauma, and most will conform to the criteria established in the DSM. In the sequence opposite, (a) an infant orphan witnesses the death of his mother and family during a cull; (b) the terrified infant is tethered to the dead bodies of his family while other orphans are gathered and tethered; (c) after the ivory, meat, and other body parts are taken from the dead elephant family, the orphans are transported in a truck for relocation or captivity. Photos courtesy of Oria and Iain Douglas-Hamilton

a
b
c

9. Jenny at the Dallas Zoo. Jenny was diagnosed with complex posttraumatic stress disorder and suffers from persistent episodes of self-injury and other severe psychological distress, including stereotypy. Photo courtesy of Les Schobert

10. The David Sheldrick Wildlife Trust has rescued and saved more than eighty orphaned elephants, a testimony to the expertise that the Trust has developed and the desperate need they fill.

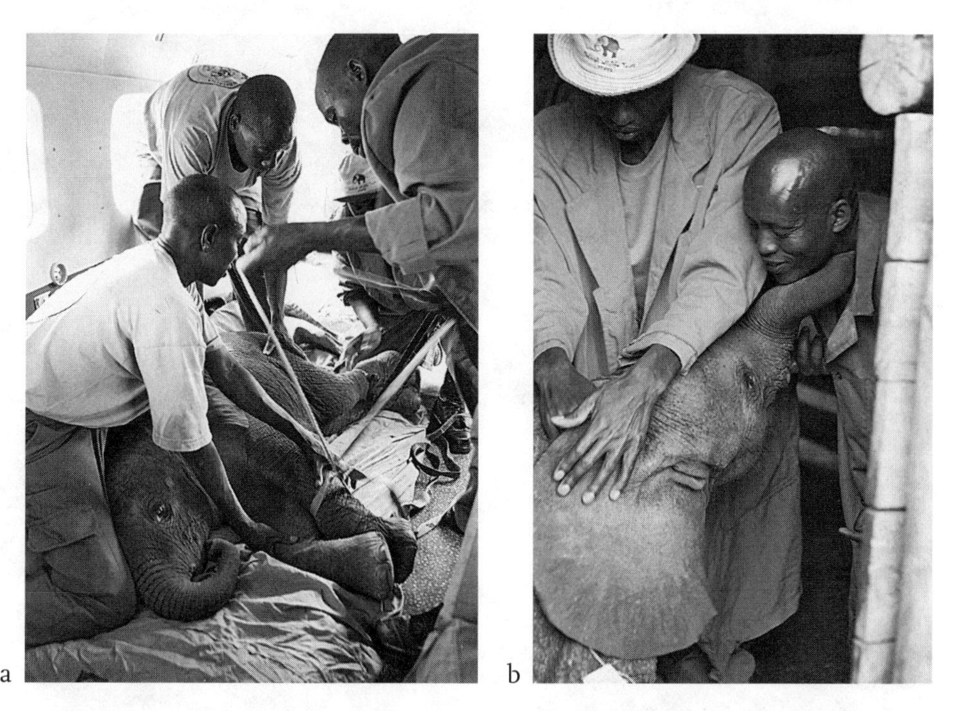

(a) Most orphans are found dehydrated and in a state of shock, and sometimes injured; all require immediate attention. (b) Saving an orphaned elephant takes precision, skill, and heart. The Trust has modeled orphan recovery after elephant culture and values complete with allomothers—in this case, the Elephant Keepers and other, older orphans who have settled into Trust life. *(Sequence continued on following pages.)*

(c–e) The bonds they form are deep and vital to their recovery.

f

g

(f) After they are about ten years old, the orphans graduate from the nursery and are gradually reintroduced to the wild.
(g) They never forget their erstwhile human family, however.
Photos courtesy of the David Sheldrick Wildlife Trust

11. One of the Sanctuary Divas, Frieda, in her former brutal circus life.

12. (a) Flora, formerly of Circus Flora, joins the Elephant Sanctuary in Tennessee, where she learns to come into her own as a mature female African elephant.

a

b

c

(b, c) At the Sanctuary, the residents live in fresh air, with running water to frolic in, barns to retire to, hills to climb, trees for shade and knocking down, nutritious food, company, and perhaps most important, freedom to do what they please, how and when they please. Captive life has taken a toll, but the Sanctuary residents are thriving and working to re-create a new life and culture.
(Sequence continued on following page.)

d

(d) Tange and Zula (who has since died), two other African elephants, with Flora. In sanctuary, the elephants are able to make friends and bond, an essential aspect of trauma recovery. Unlike zoos and circuses, the Sanctuary elephants can stay with their loved ones until death parts them.
Photos courtesy of Carol Buckley

13. (a) Although the armor and leathery skin of a tortoise seem ill-suited for massage, Elke Riesterer's "listening hands" are welcomed by Ralph the Aldabra tortoise, who is more than one hundred years old. (b) Elke's work at the David Sheldrick Wildlife Trust, where she visited the orphans and used body-centered therapy, has included training Keepers in using her techniques.
(Sequence continued on following page.)

c

(c) Dunda, now at the Oakland Zoo, receives Elke's therapeutic care.
Photos courtesy of Elke Riesterer

7

The Sorrow of the Cooking Pot

During the 50 plus years that I have been intimately involved with Elephants in Africa, and the rearing of over 80 orphans, I am astounded about how forgiving they are, bearing in mind that they are able to recollect clearly that their mother, and sometimes entire family, have perished at the hands of humans. Our Elephants arrive wanting to kill humans but eventually protect their human family out in the bush, confronting a buffalo, or shielding their surrogate human family from wild, less friendly peers. That is why I say that they are amazingly forgiving, because there can be nothing worse in life for an Elephant than witnessing the murder of those they love. And since Elephants never forget (which is a fact), they demonstrate a level of forgiveness that a human would in all likelihood have difficulty in achieving.
—Dame Daphne Sheldrick

We now explore what happens after trauma, what happens in the wake of pain "deposited in the bones" and how an individual, community, and culture recover from cataclysmic change. Trauma treatment typically focuses on those who have lived through the experience; indeed, it is their plight that moves us to compassion and inspires us to help them to heal. But victims are not the only ones involved. In keeping with attachment theory and neurosciences, Archbishop Desmond Tutu observes that neither are we apart from those who have suffered nor are we immune from their pain. "We are bound up in a delicate network of interdependence because, as we say in our African idiom, a person is a person through other persons."[1] Such interdependence extends to ethical responsibility: silence in the face of the dehumanization of one inexorably dehumanizes the other. The Martinique psychiatrist and revolutionary Frantz Fanon phrases it more sternly: "I cannot dissociate myself from the fate reserved for my brother. Every one of my acts commits me as a man. Every instance of my reticence, every instance of my cowardice, manifests the man."[2] Elephant and human suffering and their recoveries are linked.

When we consider the task of putting the pieces of a single life back together, then multiply it by the thousands or millions, the result can be overwhelming. It is difficult to imagine how an elephant can return to an earlier life, or even enjoy a present given such acute losses; contemplating the enormity of elephant breakdown and its attendant grief can lead to a kind of emotional paralysis. But the story of trauma is not hopeless. Astonishingly, traumatic experience has catalyzed profound positive transformations for some. The psychiatrist Robert Jay Lifton writes: "In almost all of my interviews with people, with Hiroshima survivors and, say, survivors of Nazi camps, as different as they are, they talk about something that they have learned; some have amazed me and troubled me by saying that they would not want to have missed that experience."[3] For those haunted by memory, this sentiment may be difficult. Recalling a moment during the weeks after the Buchenwald liberation, Elie Wiesel wrote, "One day when I was able to get up, I decided to look at myself in the mirror on the opposite wall. I had not seen myself since the ghetto.

From the depths of a mirror, a corpse was contemplating me. The look in his eyes as he gazed at me has never left me."[4] It is not easy to think of reviving emotions of hope and joy when a painful past is ever-present.

Difficulties notwithstanding, human testimonies and Dame Daphne Sheldrick's observations of elephants illustrate how renewal is possible. The elephants she has saved have not strayed down the path taken by the rhinoceros-killing young bulls. Her orphans are able to rejoin their compatriots in the wild and mature into fully functioning members of elephant society, even showing compassion for the species responsible for their predicament—humankind. They appear to be able to take the first step toward creating a new relational balance between species. If elephants have this capacity and an apparent willingness to begin anew peacefully, can we who have caused others "deep, deep anguish and pain and suffering" show remorse for what our species has done?[5] Are we able and willing to change our ways enough to break from the spiral of violence? These are some of the questions involved in elephant trauma recovery.

Most of our discussion thus far has centered on similarity—science's new appreciation of species' mental and emotional overlap, and in particular, how much people and elephants share down to the very level of how we each become who we are. Newfound kinship evokes a fresh sense of excitement after a long-held fear of anthropomorphic trespass and segregation from our natural union with other species. Finally, a sanctioned path has emerged in the language and concept of science for understanding an elephant much like we would a person from a different culture. Science has rendered species differences on the level of *cultural* differences rather than biological ones.

After seeing the neurobiological nuts and psychological bolts of elephant experience from the eyes of human traumatology, and learning how much both species share, it is time to assess what elephants and modern humans do not share. The fact that neuroethological patterns are more or less predictable across species is a fairly straightforward proposition. Similar neurobiological levers and cogs are engaged in parrots, belugas, elephants, and people when psychological shock occurs.

Witnessing death or experiencing its threat engages a primary reality; the aftermath of grief and mourning is a universal response.[6] But the comfort of similarity ends there, because a description of recovery is much more complex.

While trauma may affect every person, its specific meaning and manifestations are neither predictable nor universal. Just as differences in the ways we are raised influence who we become later in life, so do experiences of trauma. The intrinsically personal and philosophical nature of trauma makes recovery resistant to facile translation from one individual to another, or the usefulness of any off-the-shelf, one-size-fits-all formula. For some, well-being increases steadily with time. The greater the distance from the traumatic event, the better the chances of reassembling a meaningful life. For others, grief and depression come and go, rise and fall, and morph from one set of symptoms to another.

An orphaned young bull, Mzima, who eventually made his way to the Sheldrick orphanage, had been sighted for successive days near Kilaguni Serena Safari Lodge, struggling alone in the bush, obviously becoming increasingly more desperate to reunite with family; he even tried to follow a zebra herd. After rescue, and almost upon arrival at the orphanage, Mzima seemed to gleefully take to his new surroundings, showing affection to humans and playing with the other young residents. In contrast, Dika, an orphan who saw his mother slaughtered, could not be consoled for weeks on end and was so distraught that he almost perished from grief. However, Mzima's elation was short-lived; after a week, he had lost his cheer and became sad and withdrawn. He is, in Dame Daphne's words, "no longer tasked with just trying to survive, he is now able to contemplate the enormity of his loss, and this, sadly, only time will heal."[7]

Psychological trauma is definitional, permeating all aspects of life down to the bare bones of existence itself, regardless of faith or political views, and across all types of relationships, professions, and lifestyles. When "drenched" in the conditions of extreme violence, writes Edward Tick, a clinical psychologist and author of multiple books on war and American veterans, "the soul is disfigured and can become lost for life. . . . It is the removal of the center of existence from the living body without completely snapping the connection. In the presence of overwhelming life-threatening violence, the soul—the true self—flees."[8] Away from events, when the survivor is released from these circumstances, the traces of his trauma may seem to an outside eye to be gone—normal life is resumed, families are reengaged, work is performed, holidays are celebrated—but appearances can be misinterpreted. Elie Wiesel commented after hearing that a novelist friend had plummeted three stories to his death: "Primo Levi died at Auschwitz forty years later."[9] Measured by one philosophical yardstick, Levi had "recovered"—he was a successful writer, internationally acclaimed, and lived for decades after his horrendous experience in the concentration camp. By another yardstick, Levi's death was evidence that he had not recovered, that his material accomplishments and longevity had done nothing to heal the internal scars that, Wiesel implies, prompted Levi to take his own life.

Levi's story illustrates the complexity confronting both the survivor and those who witness survival. Extreme trauma propels the survivor into a universe very different from the known world, out of ordinary woes and preoccupations—traffic jams, grumpy bosses, ill-fated romantic interludes, and paying the bills—into a bewildering unknown where past and present realities are incompatibly thrown together. Life after trauma even seems to run on a different timetable. Unlike the rhythms of the school year, daylight saving time, and nine-to-five jobs, recovery knows no discrete starting point or endpoint: measure and form of recovery are relative.

Neither is recovery trauma in reverse—there is no going back. In story after story, victims of violence speak of the irrevocable change that ripples through their lives long after the event. Abducted and tortured while serving on a Catholic mission in Guatemala, Sister Dianna Ortiz writes: "No one ever fully recovers—not the one who is tortured, and not the one who tortures."[10] It is this profound change, the indelible nature of trauma, that belies a definition of recovery as a return to the prior, original state. Sister Dianna raises another key element: trauma may be personal, but it is relational at heart. Like development, injury and repair of the self take place in relationships: the recipient and agent of violence are intertwined, and so, subsequently, in recovery, are the wounded and healer.[11] Nonetheless, this connection is often denied or remains unrecognized for various reasons.

For one, traumatic experience creates a gap and discontinuity between the survivor and the ones to whom he returns. This sense of disconnection and alienation is invoked in a poem by the Vietnam veteran Nathan Marbly, "But You Weren't There":

> I've got a lot to tell
> I've been to the other side of Hell
> Where people die for nothing and there's
> A lot of pain and suffering
> Where blood flows like wine
> And you're scared all the time
>
> I don't mean to freak you out but
> This is what war's really about
> So if in the night you hear screaming
> You'll know it's me . . . dreaming.[12]

Making meaning of what happened and trying to fit something extraordinary into the ordinary is unsettling for both the individual and those around him because, for the survivor, the status quo of life-as-it-was can no longer be taken for granted. The daunting shock a survivor experiences comes from both the traumatic event itself and the dawning understanding that life before trauma was, if not an illusion, incomplete. Life pulses on around the survivor, yet for him, the present is like a skin stretched over a raw alternative reality where memory threatens to intrude. Even the most supportive family and community may never be able to bridge the gaping hole. They can only hear the screaming; they cannot access the experiences and world that Marbly revisits in horror.

Trauma introduces an intrinsic inequality. The veteran returns from war, radically changed inside, and sometimes physically, to an unchanged home. The world in which trauma occurred may be separated from the "everyday" world in time, but the two intersect in the mind and body of the survivor, something often undetected by family and friends. To the observer, the traumatic event is a discrete part of the past. For the survivor, as Marbly describes it: "The war is over in history, / But it never ended for me."

This idea—the relational and boundless nature of trauma—accounts in part for why traumatology has had such a profound impact on methods and models of modern psychiatry. In keeping with the overall mission of science, the medical model used to diagnose trauma-induced symptoms and conditions like PTSD codifies subjective experience with uniformity so that it is readily transported across gender, ethnic, racial, geographical, and other separating boundaries.[13] Whether or not we are happy with the global proliferation of science, its agenda of objectifying universality has had a huge impact on peoples and cultures as widely spaced as Finnish Lapps, Rwandan Hutu, and South Pacific Samoans. It can be thought of as a lingua franca to some degree. However, there has been a cost. The complexities and ambiguities of personal experiences and cultural traditions that provided coherence in past uncertain times are lost with the leveling effect that universality imposes.

Postmodern and modern models crashed head on when American politics first confronted PTSD and the Vietnam veterans' rights movement. Edward Tick describes the searing disconnect when returning veterans "come home stumbling out of hell." The communities to which war veterans return fail to "see them as they have become. Instead, we offer them beer and turkey dinners, debriefing, and an occasional parade, and a return to routine jobs and weekends in the shopping malls."[14] Americans' superficial attitude toward veterans—their offerings of "beer and turkey dinners" and their unwillingness to help vets make sense of the horrors they have seen—derive from a uniquely modern ailment: the dismissal of psychological suffering, the denial of death, and a consequent inability to mourn.[15] That many Americans showed outright disdain for soldiers returning from Vietnam exacerbated the veterans' ability to recover from their trauma.

Elie Wiesel encountered a similar disavowal of history when he first tried to get his book *Night* published. The book, today translated into multiple languages and sold in the millions, was "rejected by every major publisher, French and American, despite the tireless efforts of the great Catholic French writer and Nobel laureate François Mauriac. . . . After months and months of personal visits, letters and telephone calls, he finally succeeded in getting it into print." Wiesel brings up something subtler concerning the chasm between those who were there and those who weren't. Like Marbly, he speaks of disjunctive worlds: "Only those who experienced Auschwitz know what it was. Others will never know." He evokes the ineluctable tension between such experience and its awkward translation into the collective medium of language:

> I would pause at every sentence, and start all over and over again. I would conjure up other verbs, other images, other silent cries. It still was not right. But what exactly was "it"? "It" was something elusive, darkly shrouded for fear of being usurped, profaned. All the dictionary had to offer seemed meager, pale, lifeless.
>
> And yet, having lived through this experience, one could not keep silent no matter how difficult, if not impossible, it was to speak. And so I persevered. And trusted the silence that envelopes and transcends words. Knowing all the while that any one of the fields of ashes in Birkenau carries more weight than all the testimonies about Birkenau. For, despite all my attempts to articulate the unspeakable, "it" is still not right.[16]

It is for this "it-ness" that Patrick Bracken and his colleagues Joan Giller and Derek Summerfield fight against the threat that modernity poses. They define modernity as "a valorization of reason and science and a disregard for tradition, religion and disorder." In contrast to healing practices of pre–twentieth century Europe and indigenous cultures, Western psychiatry has "sought to convert the human sufferings of madness, misery, and alienation into technical problems . . . which are amenable to technical interventions."[17] Communal traditions—from tribal giveaways connecting the lives of the living with those of the dead, to Irish wakes celebrating death in life with days of communal song, drink, and oratory, to elephant rituals in which the family gathers around the dead body of a relative—function as public recognition of the inflection point between life before and after death. Many such traditions have withered. Bodies are no longer laid out, washed, and wailed over. Instead, corpses quietly disappear through back doors and funeral parlors.

Psychology and psychiatry's emphasis on the individual has also been criticized for minimizing the social nature of trauma.[18] Society's inability to feel related to what the survivor has experienced constitutes a fundamental rejection of the social contract, and the reinforcement of an already antiseptic segregation of the living from the afflicted, dying, and dead. Instead of restoring the communal life force, mourning denied transforms into an arrested melancholia in which the "would-be mourner makes the lost person into part of his or her self. The two are no longer psychologically separate from each other, if they ever were. As a result, the subject unconsciously comes under the influence of the lost person."[19] Death has failed to individuate, and unrecognized trauma seeps into the fabric of culture, with the result, Tick concludes, that "we as a nation are trapped in a consciousness that cannot acknowledge abject suffering, especially if we have caused or contributed to it; we do not see the reality of war."[20] The survivor knows that the world of those who have watched from the sidelines is not more real than his trauma world, yet frequently it is treated as such. The trauma world is something unseemly, irrelevant to the present and future, and best kept from view.

Here, we encounter something reminiscent of Hannah Arendt's "banality of evil."[21] The need to see bad deeds as something extraordinary and to see the survivor's experience of trauma as something exceptional blinds us to the potential of violence in the players of everyday life.

Since even before the Vietnam War, therapists and veterans' advocates have decried not only the inadequate material resources allocated to returning soldiers but the psychological segregation imposed on veterans. Such segregation pathologizes the survivor, the victim of violence, by refusing to acknowledge the link between the collective agenda of war and the legacy of individual despair. When society insists that the world of camps, torture, mutilation, and death exists apart from everyday life, the survivor is condemned to psychological purgatory, to remain unseen because his past has no place in the present. "Survivors feel trapped in that apocalyptic reality and rarely try to explain it to people who will not understand."[22]

From the perspective of social psychology, PTSD emerges not only as a veterans' problem; it is a disorder of identity, a "soul illness" meted out to the individual by an amnesiac collective through sanctioned violence.[23] It is the confrontational nature of recovery, its insistence that past and present, cause and effect be brought together and mutually recognized, that makes trauma political.[24] Ignacio Martín-Baró, a Spanish Jesuit priest and psychologist, founded the field of "liberation psychology" to bring attention to the highly political nature of psychology and suffering.

Martín-Baró maintained that in order to facilitate healing for the good of the individual, psychology itself, and with it the beliefs of those in power, had to be liberated from the concepts, practices, and institutions that caused the trauma in the first place. While working in El Salvador, he discovered that poverty and oppression—collectively engineered or at least facilitated—were in large part responsible for undermining mental well-being. The minds and hearts of individuals must be understood from their own points of view, not from an enforced group standard, and Martín-Baró appealed to his colleagues to acknowledge this connection. "If we want psychology to make a significant contribution to the history of our peoples . . . we have to redesign our theoretical and practical tools, but to redesign them from the standpoint of the lives of our own people: from their sufferings, their aspirations, and their struggles."[25] If we have any doubt about the politics of suffering, or about how high the stakes are, we need only recall that in 1989 the Salvadoran army murdered Martín-Baró for his revolutionary stance.

Even when politics is acknowledged, psychiatry encounters difficulties, and critics find its structure wanting. Applied outside its culture of origin, psychiatry, with its diagnostic boxes, ends up clipping off indigenous meaning for those with vastly different histories, mythos, and values. Psychiatry falters in its ability to heal. The experience of Deogratias Bagilishya, a psychologist at the Transcultural Psychiatry Clinic of the Montreal Children's Hospital, illustrates the problem.

In August 1995 Bagilishya arrived in Rwanda in search of his twenty-one-year-old Hutu son, Yves, who had been missing for a year. A practitioner himself, Bagilishya suddenly experienced the limits of modern psychological models when confronted by his son's death. The killer sent a young Tutsi soldier to deliver the devastating news. Bagilishya describes his experience of intense suffering and disorientation:

> The paralyzing revelation of the death of my son transformed my inner world into an ocean of tears and sadness, buffeted by a storm of anger that laid the foundation for thoughts of bloody vengeance. Unceasingly, my inner voice was telling me it was my right and my duty as a father to avenge my son—that one death demands another. I felt shame and guilt for these violent impulses that flooded my mind. I was frightened by the disagreeable feeling of hatred that filled my being but was incapable of distancing myself from this overwhelming hatred that demanded action.[26]

Prompted by his mother's gentle intervention, Bagilishya invited the young emissary inside. At first, he listened in his familiar role as therapist to the soldier's account of the horrors of Rwanda's implosive genocide. But then the conversation took an unexpected turn. Monologue turned into dialogue and a rhythmic exchange. The simple trade of two proverbs—*Akamarantimba kava mu muntu* ("the greatest sorrow comes from within") uttered by Bagilishya, and the young man's answer, *Agahinda kinkono kamenywa n'uwayiharuye* ("the sorrow of a cooking pot is understood by him who has scraped its bottom," signifying that one can help someone else only by genuinely listening to his suffering)—brought communion. A Rwandan tale conveyed the penetrating contagion of trauma in a way that sociological analyses could not. "A kid [goat] accidentally overhears that a man intends to slit his throat. Overwhelmed by sadness and anger, he refuses to nurse. His mother questions him and, overcome as well, she refuses to run away from a leopard. Himself possessed by their sadness and anger, the leopard sits immobile before the man who shares his own feelings with a cow."

Beyond their poetic lyricism, traditional Rwandan images and proverbs capture the broader historical and environmental significance from which personal meaning spawns. Bagilishya found that "the use of foreign therapeutic models, organized around concepts like 'post-traumatic stress disorder,' must raise questions about their pertinence and their positive and negative effects. Even if a foreign model is not necessarily harmful, it must be examined with extreme care to avoid destroying the fragile internal equilibrium."[27] Such experiences have stimulated the emergence of transcultural psychiatry, which like trans-species psychiatry, is gradually making its way into the broader field of mental health.[28] Bracketed within the rich expanse of time and space, modernity is slowly becoming a case study in lieu of its former status as an absolute frame of reference.

How then to approach elephant trauma recovery? If recovery is so tied to specific cultural values and practices, can human traumatology hope to offer any insights for reviving elephant minds and culture? And given that the elephant and modern human cultures have significant distinctions, as is evident from their divergence in behavior, are human interventions even needed? Finally, if, as in the case of veterans, elephant recovery is possible only if human society itself becomes involved, can human culture change fast enough to save the elephant?

Good questions, and challenging. However, we are able to answer them now in part. For example, in responding to the first question, we may look to what Dr. Bagilishya's experiences have shown: that much can be gained by looking at elephant recovery from the perspective of elephant cultural values and rearing practices. This is available in the detailed documentation by elephant ethologists, individuals such as Dame Daphne Sheldrick and Cynthia Moss, who are versed in elephant society and minds, and the knowledge of people who have traditionally lived with the species. To utilize principles and methods of traumatology, it is necessary "to redesign our theoretical and practical tools" to fit those in need: in this case, to design concepts of recovery and therapeutic approaches from the standpoint of elephant lives, their sufferings, their aspirations, and their struggles, and to center inquiry, as much as possible, from within the elephant psyche. Indeed, this is exactly what Dame Daphne has accomplished, which brings us to the second question.

Yes, human intervention is necessary and useful, but it must be tailored and conditioned on the specific circumstances, with an implicit caveat: "If it's not broken, don't try to fix it." Historically, elephants left to their own devices without the persistent threat of harassment and death, with access to appropriate tools of recovery, such as good food and water and ample freedom to roam their traditional lands and revitalize their social network, do very well with no human aid. But times have changed, and elephants lack many such resources because of humans. Now we arrive at the last and most difficult question: are humans, as the primary agent of elephant trauma, able to commit to the change required?

Judith Herman, Ignacio Martín-Baró, Edward Tick, and many others insist that we must look beyond what is already known to be the proximal causes of trauma—in the case of elephants, capture, captivity, brutality, and mass killings—to understand how to prevent future suffering. The experiences of veterans, political prisoners, and victims of domestic violence make clear that the roots of systemic violence extend deep, beyond action, to individual and collective psychologies and values responsible for forging the instruments of trauma. Separating the survivor from the agent of trauma requires dismantling the attitudes, the perceptions, the relationships, and the institutions that facilitate the proliferation of a single act of violence into millions. In other words, to address elephant breakdown, we must accept that elephants, and other animals, have rights comparable to those of humans. To grapple with recovery is to grapple with the basics of existence itself—to ask how to live.

This is tricky ground. We are treading ever closer to the sensitive area of human privilege. Speaking about the comparability of elephants and humans in scientific abstractions is one thing, but putting ideas into practice is another. Fear of anthropomorphizing, falsely attributing human characteristics to other animals—or, interpreted more psychologically, the fear of being like animals—has been ingrained in scientific and social training. That fear has helped to maintain the barrier between us and elephants and other creatures. But traumatology, science itself, tells us that in order for elephants to live, then it is we who must change. Elephants' recovery is dependent on human capacity to change. Instead of Happy, Jenny, Black Diamond, and their compatriots, modern psychology and culture recline on Freud's famous couch as we rediscover what has been lost in the frantic race of progress, to learn how to embrace elephant ways to create a truly kinder and gentler way of existence.

8
The Biology of Forgiveness

Girija Prasad, also known as Manikantan, is a twenty-year-old Asian elephant who worked for many years as a temple elephant in India. In 2004 he was rescued by Compassion Unlimited Plus Action (CUPA), an animal welfare organization based in Bangalore, after a police complaint against Girija's treatment was filed. Girija was brought to the Bannerghatta Biological Park. Dr. Surendra Varma of the Asian Elephant Specialist Group (AESG) evaluated Girija's condition as dangerously poor, and Girija was diagnosed with complex posttraumatic stress disorder, common among long-term prisoners. More than sixty wounds to his face and neck alone were identified. On March 13, 2008, after four years of negotiations and multiple hearings in the Karnataka High Court, Girija won his case. In a precedent-setting move, Chief Wildlife Warden I. B. Srivastava ordered that the temple's owner

certificate for Girija be cancelled. This was the first time in India that a temple's ownership of an elephant had been revoked because of ill treatment. CUPA is currently petitioning the Supreme Court to define elephant cruelty in legal terms as it applies in future cases of abuse.

Recovery seems to lack any obvious polestar, no clear path forward when the past remains close at heel, its terrors breathing heavily in the survivor's ears.[1] Traumatology provides theories, but there is no promise of what the future entails, and there is an unsettling suspicion that wrongs do not get righted, no lessons have been learned, and that all has been for naught. Yet there are those who somehow breathe meaning into life, who serve as reminder that while restitution may not come within a lifetime, the past can be prevented from becoming the future. Dr. Bagilishya found a way, and Dame Daphne, through her rescue of elephants.

We have heard about her successes, and they provide reason for hope. But the seeming ease with which Dame Daphne brings infants from near death to vital teenagehood can give a false impression and obscure how hard her work really is. Statistics attest to the profound task of raising an elephant infant. Elephant births at Dickerson Park Zoo, for example, show the consequences that may transpire:

1. DEAD: Unnamed—born/died June 1985
2. DEAD: Maiya—born July 1991, died February 1993
3. DEAD: Unnamed—born/died September 1992
4. ALIVE: Asha—born February 1995, transferred to Oklahoma in December 1998
5. ALIVE: Chandra—born July 1996, transferred to Oklahoma in December 1998
6. DEAD: Unnamed—born/died September 1997
7. DEAD: Kala—born May 1998, transferred to Vallejo in May 2000, died November 2000

8. DEAD: Haji—born November 1999, died June 2002
9. DEAD: Pete—born/died April 2000 (lived one day)
10. DEAD: Nisha—born July 18, 2006, died December 2007[2]

Dame Daphne has an even greater task than zoos or circuses, because the infant elephants who come to her door arrive unexpectedly, in severe states of shock, often physically wounded, yet they do not acquire the fearful aggression and other symptoms that seem to haunt captive-born elephants. Instead, she describes them as forgiving.

Formally, forgiveness is defined as "a suite of prosocial motivational changes that occurs after a person has incurred a transgression."[3] In other words, someone who has suffered insult shows attitudes and actions reflecting a desire to maintain (or renew) a relationship with whoever caused his suffering. Generally speaking, a person who has forgiven another shows concern that is less focused on his own distress than it is oriented to others. These individuals also tend to have a strong sense of belonging, a desire for social cohesion and an extended sense of kinship. Forgiveness may reflect a change of heart, from negative emotions to positive emotions, similar to what Bagilishya found "gave meaning to our rage and sadness—and prevented behaviors akin to those of which we had spoken."[4] It thereby diffuses the impetus for revengeful reaction. It is not, however, the same as reconciliation. Reconciliation requires mutual participation, whereas forgiveness requires only one person. Much like the relationship between trauma and recovery, the act of forgiveness is asymmetric; it emanates from the wronged person, who may hope to encourage, but cannot compel, the wrong-doer to change his or her future behavior.

Despite its positive connotations and prevalence as a concept in many religions, forgiveness is not always considered positive. Some maintain that forgiveness is useful in certain cultures but of limited value in others. Others question whether forgiveness is really a possible, or even a desirable, response to horrendous crimes. And some are always skeptical whether forgiveness is an honest expression of the injured person's psyche. Is forgiveness a profound spiritual achievement, a type of dissociative coping strategy, or, as for Prospero in Shakespeare's *The Tempest*, an emotion possible only after the transgressor has paid a price?

Some of these same themes lay at the heart of the research by the anthropologist Nancy Scheper-Hughes in postapartheid South Africa in 1993. The question of how suffering, identity, personhood, and recovery alchemically express was brought to focus the day after her arrival, when three young gunmen burst into the Kenilworth Christian Church and killed thirteen people, injuring tens more. In the following months, Scheper-Hughes interviewed survivors to understand: "How did people emerging from a past characterized by competing and contradictory histories begin to build a new sense of nationhood and belonging based on a negotiated version of that same history?"[5] How were they feeling and living; how did forgiveness, remorse, and revenge emerge in the mix of race, ethnicity, suffering, and survival?

Elephant society has the same challenge. It is riddled with violence, has had ethnic (in terms of trans-species science) strife, has lost elders and the coherence of traditions, and suffers with no hope for restitution. Elephant society is also embedded in human cultures struggling with similar challenges and histories. In the vacuum of such losses elephants and humans are pressed to mobilize the remnants of their culture. Neither elephant nor indigenous human cultures can, in the immediate future, access the former wealth of social and ecological resources that fed and shaped life and offered a buffer against assault. Elephant homelands are crisscrossed with dwellings and structures, and the law of the land lies in the hands of modern humans whose values are still steeped in an agenda of possession. Even free-ranging elephants, including the Sheldrick graduates who move into Tsavo herds, must negotiate with humanity in order to survive. They and humans are asked to re-create culture, to redefine meaning in new contexts. Forgiveness, in its broadest definition, represents a pledge to the future.

Not long ago, Daphne Sheldrick's accounts of elephant forgiveness would have been shrugged off as anthropomorphic hyperbole: a wild ascription of human psychological qualities to the behavior of an animal.

Even those who have observed elephant grief might find the notion of elephant forgiveness a stretch. Grief can be intuited from these fairly accessible behaviors, but forgiveness is complex and murky. Furthermore, elephants lack the ability to use words and images employed by humans to communicate feelings and states of mind, an ability considered key to psychotherapy and recovery. So how to understand what is meant by elephant forgiveness? Is forgiveness "natural" and intrinsic to all?

These questions bring us back once more to who we are and how we develop. As we saw earlier, self-development depends on the existence and quality of social interactions. Gordon Gallup, of mirror self-recognition (MSR) test fame, found that chimpanzees who had been reared in isolation scored much lower than those raised under socially normative conditions, a finding consonant with those of John Bowlby and a string of successors who have documented how an unhappy and insecure childhood undermines a positive and healthy sense of self.[6]

It turns out that while forgiveness may retain its somewhat mysterious air, its rudimentary mechanisms can be related to neuropsychological models of self-function: forgiveness as a relational process of transformation has common psychobiological mechanisms across species. The neurobiological roots of forgiveness are evolutionarily intertwined in the functions of social living. In fact, the researcher Tom Farrow and his colleagues have even identified areas in the brain that they say recover in posttrauma therapy and revitalize prosocial, forgiving feelings and behaviors.[7] Thus despite the ambiguity of forgiveness in human culture, neuropsychology provides a firm footing and a link with something familiar and essential, the psychobiological self, and in so doing permits an exploration into elephant forgiveness.

Forgiveness requires three steps, each of which presupposes a sense of self. First the injured party must perceive a transgression against himself. Second, the party who has sustained injury must recognize a preexisting relationship with the transgressor. Third, the injured party must initiate a process of self-repair, a perceptual, emotional, and behavioral readjustment rendered through the medium of the relationship. In the language of attachment theory, these three steps describe the core neuropsychological processes of self-development, self-injury, and self-repair.[8]

Much like self-resilience, the capacity for prosocial behavior relates to childhood experience of family attitudes. Abuse, neglect, and traumatic attachment lead to a loss of ability to regulate intense feelings, producing asocial behaviors.[9] If, however, as the psychiatrist Henry Krystal noted on reflecting on those who had managed to survive the camps, the nascent self has been reared warmly and lovingly, then even when confronted with trauma, the individual is able to access the supportive caregivers and values that have been internalized unconsciously in childhood and that "permit the restoration of one's capacity for love." Those who were able to maintain familial or other ties could preserve a viable image of self to weather the obliterating power of oppression and abuse.[10] But a subtle point harks back to previous discussions relating to cultural differences, and specifically to elephant social ethology. Not only are there differences in the ways in which each individual within a culture is raised, but differences in the configuration of the self arise *between* cultures.

Most psychology, including attachment theory, has focused on the mother-and-infant bond and dynamics within the nuclear family, a social model that has almost become a caricature of the 1950s United States. However, such pioneer theorists as John Bowlby and Mary Ainsworth were aware of cultural differences and their not-so-subtle effects on rearing styles. Bowlby's three-volume work is full of examples of animal societies that depict a range of social behaviors and relationships—possessive monkeys, volatile chimpanzees, and devoted geese—leading him to conclude that "it is as necessary to consider the environment in which each individual develops as the genetic potentials with which he is endowed."[11] In other words, not everyone grows up in television's Cleaver family.

One glance around the multicultural society is evidence enough of the rainbow of family life. TV situation comedies and dramas like *All in the Family, The Cosby Show, The Sopranos,* and even *The Addams Family* celebrate this diversity of ancestry and family. Statistically, when the sum of human cultures is considered over history, the modern Western style of raising children is more the exception than the rule. The psychologist Patricia Greenfield and other transcultural researchers have taken note of this fact and extended the rudiments of attachment theory to include the multiplicities of rearing styles and values across cultures.[12]

According to Greenfield and others, child rearing in human cultures can be grouped into two broad categories. One category, characteristic of modern Western, Anglo-American cultures, is based on values of individualism and independence. Child rearing is reflective of the self and personhood resonant with St. Augustine's model of a self who is interactive with but separate from those around him. Individual autonomy and a well-defined distinction between self and other are stressed. From Sigmund Freud to Dr. Benjamin Spock, doctors and psychologists have taught that successful mothering involves implementing a succession of separations and reunions with the infant in order to cultivate a child who is mentally and emotionally differentiated from his or her parents. Separation from nature and from mother (from Freud's point of view, the most important separation) is vital for healthy psychological maturation. This typology of self is agentic (capable of taking initiative), rationalistic, and univocal.[13]

In contrast, many non-European indigenous cultures do not consider separation of mother and infant essential to good child rearing. Examples from Cameroon, rural India, Mexico, and elsewhere illustrate an alternative "null" cultural model of child-rearing practices, one that is based in values of collectivity and interdependence. Instead of cultivating the process of separation and reunion, caregivers keep a baby in much closer contact, day and night, and give less emphasis to enforcing the infant's self-other distinction. Interdependence is reinforced in other ways. The mother may be the primary source of milk and care, yet her infant, much as we saw for young elephants, will often spend more time with other relatives.

Australian aboriginal infants are "knee babies," nursed and raised by multiple women in the community other than the birth mother. Generally, weaning comes later for these children, at three to five years of age, a practice that is not only discouraged but considered psychologically dubious in conventional Anglo-American cultures.[14] Aboriginal rearing and other collective, interdependent cultural contexts more closely resemble a constellation than the single star-mother around which a baby revolves when raised in individualistic, independent cultures. Furthermore, not all cultures are dominated by the "eye-to-eye" or "face-to-face" bonding characteristic of Euro-American cuddling, which emphasizes the self-other distinction.[15] Indeed, direct eye contact is considered offensive and confrontational in many American Indian cultures, such as the Navajo.

All of these variations translate neuropsychologically.[16] A stern aunt, a playful uncle, a concerned grandmother, or a melancholy cousin, each uniquely contributes to the tuning of the infant mind, emotions, and identity, and because attachment processes are so intimately linked with the genesis of the self, we would also expect to see culturally contingent models of self—which we do. Again, contrary to modern Western culture, where self is individualistic, autonomous, and also anthropocentric, ecocentric, and cosmocentric, cultures of Australia's aboriginal tribes extend their concepts of interdependence to relational transactions with the nonhuman worlds. Inuit personhood is created through and defined by relationships with native animals, as well as with water, sky, and earth. Aboriginal children become intimate with their natural surroundings and are initiated into the secrets of the plants and animals that create a communal and ecological, and spiritual identity.[17]

Accordingly, the concept and value of forgiveness also vary with culture. A study contrasting cross-cultural concepts of forgiveness in Congolese and French communities found that what motivates forgiveness relates directly to fundamental cultural values and representations of self.[18] By and large, the ways a Congolese and a French national perceive the meaning and method of forgiveness reflect their respective concepts of self. For collectivist, interdependent cultures like the Congolese, the way an individual perceives and responds to a transgression committed against him is somewhat "diffuse"—that is, there is a less individualized sense of transgression. Because the Congolese self is more collective, an aggregate of multiple individuals whose identities are dispersed among group members, the perception of a transgression is perceived accordingly. In contrast, the researchers found that forgiveness in an individualistic, independent culture like the French tends to be more reliant on intrapsychic properties. To forgive or not to forgive is less a collectivized concept than the offended individual's own choice. In relative terms, individuals from collectivist Congolese communities were more apt to let their resentment dissolve and to embrace reconciliation than were the French.

Translating concepts across cultures and languages is always uncertain and making generalizations is even more so. However, there is enough evidence from neuroscience, psychology, cross-cultural studies, and anthropology that tendencies toward prosocial behavior leading to forgiveness can be significantly influenced by the cultural values and beliefs in which a person is raised. Casting an eye back to elephants, it is clear that their culture rests on collective, interdependent, prosocial values and practices. Birth, life, and death are all communal in elephant society, and it may be that because their trunks provide greater flexibility in interpersonal contact, elephant attachment processes do not rely on face-to-face, eye-to-eye contact alone, instead favoring an all-encompassing somatic, acoustic, and pheromonal exchange.

Elephant society might very well have served as a model and inspiration for Hillary Clinton's mantra that "it takes a village to raise a child." While the mother remains focal for nursing, during the first four-plus years (when brain and behavior are the most plastic), the infant spends the majority of time in the company of the collective natal family. The free-ranging elephant self is defined through relationships; infant elephant cognition, emotion, behavior, and values are created in plurality. A prosocial self is the by-product of such processes: an elephant self that is group oriented and, we might go so far as to say, defined by multiple relationships.

What goes up, must come down. In theory and according to the testimony of Deogratias Bagilishya, the self in traumatic disrepair profits from taking place in a context much like its developmental origins. In order for the Sheldrick orphans to recover, to be able to survive with, integrate back into, and be functional in the Tsavo elephant communities, they must acquire the values, customs, and social contracts of elephants. But first they must survive the initial shock of trauma. One Trust orphan, Imenti, now grown and reintegrated with the wild herds, provides a vivid example of these ordeals. Imenti came to the orphanage still covered with fetal membrane. He had been literally born into trauma when his mother gave birth to him while being hacked to death by irate villagers.

Dame Daphne was not always able to save those as young as Imenti. Few younger than two years old survived. The biggest problem was providing appropriate milk. Finally, in 1974, when Dame Daphne developed and perfected a milk formula, the infant mortality rate dropped significantly. Her discovery has been lauded by veterinarians and elephant experts the world over. This pediatric advance represents more than technical progress. It reflects the careful attention that Dame Daphne has devoted to understanding details of elephant life and culture. She is now able to rear an elephant from just hours old to the teens. From carefully created formula comes carefully designed nursing.

Ndume and Malaika are two elephants who came to the orphanage after they were separated from their families and who, like Imenti, almost died at the hands of angry farmers. Armed villagers had attacked the elephant family, who had wandered from their tiny fragment of forest into cropland. When dawn broke one morning, the elephant family was spotted standing in the midst of a maize field. The villagers rushed toward them, throwing spears, firing poisoned arrows, and hacking into flesh as the panicked elephants fell to the ground. A few were able to break free and flee back to their forest to relative safety, but three infant calves remained alone, Ndume, Malaika, and a smaller calf who was literally cut into pieces before them.[19] The Kenya Wildlife Service Rangers intervened, and the two three-month-old elephants who were still alive were brought to the orphanage to be treated for shock and *panga* (knife) gashes. The young survivors managed to live through the first hours, always a crucial time, and gradually began their day-by-day fight to recover.

Much to Dame Daphne's surprise, Malaika, who had been "almost hamstrung by panga blows," appeared "unusually happy, obviously relieved to find humans that were not bent on harming her." Her survival seemed assured. On the other hand, the prognosis for Ndume was far less bright. A serious head injury had rendered him unconscious, necessitating an intravenous saline drip to an ear vein. Sheldrick recalls that when Ndume regained consciousness and revived, he began to "cry pathetically and bellow for his mother. At night, he had difficulty sleeping and often we would wake to hear his screaming and shrill trumpeting—a distress signal of a baby—obviously reliving the traumatic events in his dreams."[20]

Meanwhile, Malaika's initial euphoria was short-lived, and what promised to be a speedy recovery reversed. She sank into a deep depression that lasted for weeks. Dame Daphne recounts:

> It was terribly upsetting to witness such distress—the fruitless search of Ndume for his elephant family, and the depression suffered by his little friend. The two baby elephants were a tragic sight, as lifeless as zombies. The hours of darkness were filled with bellows because both babies suffered nightmares. Ndume's bellowing became so frantic that we knew we had to let him out of his stable so that he could be reassured that his mother was no longer around. As soon as he was out he tore about frantically searching for the mother he would never find until the Keepers gently rounded him up, and brought him back to Malaika.[21]

After Ndume's frantic search for his lost family ceased, a profound grief set in. Eventually, he appeared to despair of finding his family, and in the evening he began to hungrily nurse from his bottle and "snuggle close" to the Keeper. Then, little by little, both orphans settled in and began to take interest in other young, recovering orphans, and the world around them. Interacting with other elephants worked like an elixir. As Sheldrick reflects, "Elephants possess compassion in abundance, and this helped the healing process. It was, indeed, a joyful sight when they began to play, for we knew then that we were winning."[22]

Every step of the orphans' care has direct influence on their survival. From the moment they arrive at sanctuary, the traumatized elephants require the constant presence of caregivers. While Dame Daphne began this work with only her husband and daughters, there is now a constellation of Elephant Keepers who staff the critical care, nursery, and reintroduction teams.

These Kenyan men have been taught every aspect of the elephant's life. Keepers must attend to the orphans twenty-four hours a day, whether watching them on the monitor, coddling them, or feeding them. Every night each orphan, bedded in straw and covered by soft blankets, sleeps next to a Keeper. With the break of day, they join the other infants and nannies to feed, play, and do the most elephantine of things—socialize.

As the infants age, the Keepers, aided by older elephant orphans, step into the shoes of allomothers and teach the youngsters what is edible and which grasses are succulent, as well as what constitutes good and naughty elephant behavior. The knowledge that the orphans acquire forms the basis for their elephant education. *Acquire* is a key word here, because it reminds us that what elephants know does not all depend on genes or "instinct."

Earlier, we learned of Billy Jo, the human-reared chimpanzee who almost died because of lack of chimpanzee know-how. His clinical diagnosis of major depression related to a psychological wound that refused to heal, the simultaneous inability to be part either of chimpanzee society or of what he considered to be his natural, native culture, human society. By seeking to mirror natural natal family-group processes and elephant cultural values, the Sheldrick Trust creates a trans-species culture that promotes social behavior, social cohesion, and an orientation toward the group resonant with Tsavo elephant ways. In John Bowlby's language, the Trust's Keepers provide positive attachment figures, nurturing human allomothers whose behaviors and methods critically match elephant culture and biology: every system of psychotherapy depends on cultural models of the self.

Later, during outings in the bush with the Keepers, the orphans have contact with wild elephants and initiate their gradual exposure to a broader community and future. This is a time when they can "test" what they have learned in the nursery and come into their own to the limits of their Keepers' apron strings. The young ones walk with Keepers far and wide in the bush, encountering wild elephants, but always rejoining their human family in the night stockade for protection against predators to whom the young pachyderms are still vulnerable. As they approach their teens, the orphans are evaluated for emotional and psychological preparedness. The move to Tsavo is made as trouble-free as possible; it includes an extended transition period during which the newcomers mingle with humans and older Trust orphans. If deemed self-sufficient, they leave the Trust nursery and are reintroduced into the wild herds; they unite with other orphans whom they have met at the Trust, and eventually they conceive and rear their own infants in Tsavo National Park.

The growth from swaddling pachyderm to self-composed adult obviously works. These elephants have shown no undue aggression toward humans or other animals, and perhaps even more telling than the statistics of orphans who have survived are their individual stories. It is not uncommon for mature orphans to return, babies in tow, to visit their erstwhile human family. It is an impressive sight—a family reunion of awesome proportions.

Reunions are joyous and moving as rumbles and trumpets provide the soundtrack for trunk caresses, smiles, embraces, and laughter of old friends and family. Other graduates never return, sharing just a single farewell, one final gesture of love and appreciation, before leaving forever for unknown lives. As bittersweet as farewells are, they represent a victory for the Trust caregivers. They have succeeded in bringing an elephant soul back to life, and cultivating an elephant who feels like an elephant and is able to live like an elephant on her own with her taxonomic kin.

The transition from human family to elephant family is not always so easy, though. Some orphans cannot bear the separation from their human family. Sheldrick catalogues diverse elephant-human reunions. Imenti, who arrived enmeshed in fetal membrane, retained a powerful fidelity to his human family even after reintroduction to Tsavo wild elephant herds. He was

> desperate for his human family . . . [and] became a problem at Kilaguni Lodge until his human family joined him and escorted him back home, a journey that took them 5 days. . . . Eleanor: aged 42 instantly recognized the Keeper who cared for her when she was only 5 and whom she knew after a separation of 37 years. Uaso . . . a poaching victim . . . now often returns along with Lissa and her babies to mingle with the orphans and their Keepers. . . . Dika orphaned by poachers and reared through the Nursery from the age of 3 months [and] successfully integrated into the wild community by the age of 10 . . . returned after an absence of several years for a wire snare to be removed from his hind leg by his Keepers. He merely walked up to the Stockades, and stood stock still while the Keepers cut out the steel cable that had dug deep into his flesh, never flinching. . . . Ndume [was] raised in the Nursery from the age of 3 months, [and he is] . . . now successfully integrated into the wild community. Along with Imenti and Lewa, Ndume was relocated to Tsavo West National Park having accompanied wild bulls beyond the Park boundary into Voi town. Three weeks later, he turned up back home, a distance of 100 miles, despite the fact that he had never before been to Tsavo West, and was sedated and crated for the move. . . . Edo, an Amboseli elephant from the EB Study Group, born in 1989 and orphaned when his mother died as a result of litter poisoning. Raised through the Nursery from the age of 6 months and now comfortable with the wild herds. He returns periodically to keep in touch with the other orphans and the Keepers. Lissa has . . . had two wild born female calves, the first born in January 1999 and the second in November 2002. She and her two babies maintain regular contact with our orphans and their Keepers.[23]

The Keepers also provide a type of psychological inoculation against future trauma. The internalized image of the prosocial, multivalent human-elephant family they experience at the Trust will be a well-spring of support from which they can draw in times of stress or when they are raising their own families. In a retrospective on Holocaust survivors, Krystal concludes that "the essential attribute that permitted survival, the continuation of minimal essential functions, prevention of traumatic surrender and psychogenic death . . . and successful resumption of normal life" was early secure attachment experience.[24] Not only has Dame Daphne been able to re-create elephant rearing in sufficient detail to avoid situations like the distorted socialization of Billy Jo, the human-reared chimpanzee, but she encourages the development of a sense of self in each orphan that is both unique and distinctly pachyderm.

Contrast this evidence with conventional conservation methods, under which elephants typically are haphazardly relocated, mostly without any consideration for fidelity to land and kin. Contrast Sheldrick's experience also with the belief held by many ethologists that human contact can only be deleterious to species survival, or that such intimate contact jeopardizes the animal's survival abilities. Not true at the Trust, because elephant-human contact there is not haphazard. Even though trans-species bonds are encouraged, human cultural values are not imposed on the delicate network of an infant elephant's psyche. Nor do Trust alumni exhibit the strange and violent behaviors that occur in so many other locales where elephants are orphaned.

By definition, both sets of orphans have suffered the traumatic loss of their mothers and families. But whereas the Sheldrick orphans recover from their loss in a sheltered, loving setting crafted to look and function like the prosocial elephant society from which they came and which still exists in places, the young bulls were translocated to unfamiliar landscapes, with none of their relatives and no natal family group. It is unclear how much positive interaction the young bulls had with the older females residents at Pilanesberg. Compared with the upbringing they would have experienced before the cull, the young bulls were essentially left to their own devices. As Bowlby noted in his study of children who had suffered separation from their primary caregivers, such precarious and disrupted attachments often lead to psychological and physiological disorders.[25]

These young bulls progressed, of course, from antisocial behavior toward other elephants to the hyperaggression that resulted in so many rhinoceros fatalities. They and others who lacked the stability and protection of the Trust cultural triage have shown marked asocial behavior; many have neglected their young, and the population in Addo has suffered elevated intraspecific male-on-male fatalities, as well as aggression directed against humans.

But before moving on, it is important to retain a sense of caution before making inferences about what causes certain behavior; we must not oversimplify or confuse behavior arising from one type of motivation with another. Every behavior has a reason, or several, particularly so in the wild, where energy expenditure can be costly and interactions can be injurious or fatal. Symptoms of hyperaggression associated with PTSD may relate to but differ from actively vengeful behavior. The behavior of the young bulls who assaulted rhinos was symptomatic of their known past trauma, a type of emotional and neuroendocrinal dysregulation in which normal perception and processing of information were disabled. The young bulls' ability to act like typical elephants was somewhat scrambled. Their diagnosis of PTSD is conservative given the weight of theory and data in support. However, in other cases, a simple diagnosis may not be so straightforward.

The rhinoceros killings have been contrasted with what researchers in Kenya have called revenge: vengeful, premeditative killing by elephants of Maasai cows. We have encountered this sort of behavior before, notably in the story of Black Diamond, who waited years for the opportunity to attack his trainer. Ascribing revenge implies a sense of entitlement, defined as a normal defense against feelings of powerlessness and helplessness.[26] An entitled person "has an appropriate, reality-based assessment of the compensation to which he or she is entitled for a disappointment that can be explained developmentally as unresolved childhood omnipotence."[27] Perhaps the most famous literary rationale for revenge is Shylock's comparison of Jew and Christian in *The Merchant of Venice:* "If you prick us, do we not bleed? if you tickle us, do we not laugh? if you poison us, do we not die? and if you wrong us, shall we not revenge? If we are like you in the rest, we will resemble you in that." The argument made by Shakespeare's character is similar to our comparison between human and elephant behavior. Like Jews and Gentiles, both species, despite their cultural differences, are vulnerable to the same injuries. However, the similarity may break down here because we may not know whether or to what degree elephant cultural values historically included revenge. Neuropsychologists maintain that prosocial, forgiving behavior has evolutionary advantages for highly social species, because it helps animals to maintain good relationships, and thus increase their chances of survival.[28] Forgiveness and forgoing revenge have been linked to the neuroevolution of sociality, which implies that it is in the nature of elephants to be prosocial. Then we might ask, are cases of "elephant revenge" more recent behaviors, or at least has the frequency with which they are observed increased because of the extent and nature of postcolonial human pressures?

The answer may be a tentative "yes." Studies on human forgiveness suggest that forgiving people are less inclined to "ruminate" over past transgressions. Humans living in controlled urban environments, where food can be gotten out of the fridge when needed, room temperatures can be regulated with a flip of the switch, and doors can be locked to protect against enemies or the pesky neighbor next door, can afford to revisit the past. But dwelling on past transgressions distracts from the present and from an awareness of very real in-the-moment threats and predators; it can pose a hazard even for elephants, who, despite their size, need to be on guard and fully engaged in present-tense survival. It may be, then, that the female cow-killing elephants in Kenya are acting vengefully for a combination of reasons that make such behavior "worthwhile." They are said to have suffered from villagers' actions, so their response may reflect a stress-charged execution of a careful, calculated plan aimed at preempting further attacks by villagers.

Evolutionary theory predicts that the extensive hostility and persistent threat that humanity poses today may render prosociality and forgiveness more of a liability than an advantage. Social considerations may lose their utility and meaning in societies that are broken up and under chronic stress and threat. The Western individualistic, competitive model may serve as a more effective strategy for elephants and other animal cultures and may supplant other cultural values.

All these details and hypotheses deserve consideration, as they may help us understand how elephant recovery should proceed. We have seen the way in which the orphans rebound to elephant ways, but if stress-related behaviors are transmitted across generations, and if the young are unaided by restorative work such as the Trust performs, theory and data predict that elephant culture will change. For all their monumental work, Dame Daphne and her Keepers are a drop in the large bucket of elephant culture and its crisis. Environmental alterations are so fast and so devastating, resources such as the Trust so limited, that the future of elephant society is uncertain.

The question of elephant forgiveness also gives us cause to reflect on what may be learned about human psychology and theory. If indeed, as neuroscience suggests, we are moving toward trans-species models of psychiatry, then what we understand about our own psychology may be in for a change as well.

9
Am I an Elephant?

Week 3: The Divas are still running around like crazy—they are nonstop play activity. They have pushed over huge trees and are having a blast dragging them into pond to play. At one point there were six elephants playing in the pond together, frenetically—the water churned up like there were a bunch of piranhas under the surface. Suddenly something happened: Debbie uttered a sound and Minnie reacted. Minnie hit Debbie across the head with her trunk. Minnie slams into Debbie, one time knocking her feet out from under her, and grabs Debbie's tail, ears, and sometimes trunk. It appears that Minnie is trying to inflict pain in the same body parts that I have observed trainers inflicting pain in elephants.

Week 4: Minnie appears depressed. It is as if she no longer feels any pleasure. Her face is blank as she walks into the woods to explore. She eats with the others but does not appear to be engaging

with them. She shows no signs of pleasure, not even when she gets a watermelon. Her play is solitary and frantic, and it always includes an object that she can stomp on and break. The same thing appears to be going on with Lottie—she appears to be deeply depressed and not making any progress. She is wandering off on her own and not engaging with the others. She is displaying no joy, no play behavior. When I pet her, she does not talk or engage, her eyes do not meet mine; she looks very sad. She seems deeply depressed.[1]

One hundred and fifty million years ago, eastern North America was covered by sparkling blue oceans in whose waters strange sea creatures such as the Nautilus-shaped ammonite, the carnivorous marine reptile plesiosaur, and the alligator-like mesosaur danced and swam. When the seas eventually receded, casts of these animals were left embedded in sediments that have become Appalachian mountains and hills. Today, ancient mesosaur fossils lie beneath the padding feet of elephants in Namibia—and Tennessee. Indeed, Tennessee boasts the largest elephant sanctuary in North America, not because of any land bridge connecting continents (although imagining elephant migrations sweeping across continents, shoulder to shoulder with revived bison herds, is an appealing fantasy), but because of the Elephant Sanctuary in Tennessee.

The nearly three thousand–acre retirement home for African and Asian elephants was cofounded in 1995 by Carol Buckley and Scott Blais. Much like the Sheldrick Trust, the Sanctuary is dedicated to the healing of elephants. While the Trust must contend with vulnerabilities of bruised young minds and bodies arriving at the orphanage, they have one thing on their side that the Tennessee residents lack: time. Young elephant minds are still receptive with the promise of years of learning. The Tennessee arrivals are not so lucky. Elephants coming to Buckley and Blais's Sanctuary are typically in their thirties or forties, and they have spent years of captive life in circuses and zoos.

Through her extensive work and cumulative studies on concentration camp survivors, veterans and prisoners of war, and victims of domestic violence, Judith Herman concluded that "the diagnosis of 'posttraumatic stress disorder' . . . does not fit accurately enough . . . survivors of prolonged, repeated trauma." Rather, "the syndrome that follows . . . [such experiences] needs its own name. I propose to call it 'complex posttraumatic stress disorder.'"[2]

PTSD and complex PTSD are not unrelated: they are part of a comprehensive classification system in which trauma and symptoms range along a continuum. At one end of the continuum are individuals who have had healthy, loving childhoods and who exhibit normal capacities to deal with stress but are suddenly confronted by a single traumatic incident, such as a car accident. A parallel for elephants might be drawn in the case of a free-ranging adult who one day is speared by a Maasai, or who falls down into a pit, but recovers. The incident is traumatizing, and the individual may exhibit symptoms of PTSD, but attachment theory and traumatology predict that the experience will not overwhelm the victim psychologically the way trauma tends to affect children who are subjected to long-term abuse. As Dame Daphne's orphans illustrate, a secure attachment developed in childhood and a posttrauma social network do much to aid recovery. There is another factor that differentiates various traumatic experiences: the type of event that causes the trauma.

There are incidents that, while traumatic, leave less-lasting legacies because of how they are perceived. For example, natural disasters such as wildfire, or even the terror of a hurled spear, are probably part of elephant socioecological knowledge, what we would refer to in human terms as a social and cultural narrative. The fact that the agents of trauma are known entities, phenomena to which elephants may have adapted and may even anticipate, implies that the survivor's experience can be meaningfully shared with others in the community who may have had similar experiences. Communal sharing diminishes feelings of isolation that often come with trauma and helps maintain psychological coherence, both of which facilitate recovery.

At the other end of the trauma continuum are prolonged or multiple, highly invasive, physical and psychological insults. Victims of such trauma include abused children, prisoners, and others—including elephants in captivity—who are unable to escape their circumstances and who typically develop more complicated and enduring symptoms.[3]

Herman also created the category of complex PTSD to bring attention to the profound effects that captivity imposes on the prisoner. The effects of forced confinement have not always been appreciated, even in humans. There has been a "propensity to fault the character of the victim . . . even in the case of politically organized mass murder . . . [as in] the aftermath of the Holocaust [when there was] a protracted debate regarding the 'passivity' of the Jews and their 'complicity' in their fate."[4] Similar assertions have been made concerning victims of domestic violence, but as Herman stresses, complicity can exist only under conditions of free will, not under forced confinement. By definition, captivity relocates an individual into a completely different environment and in so doing creates a second reality that operates by different rules. Day-to-day survival plays out across the coordinates of the captive world; indeed, it *must* do so because the prisoner's life is completely in the hands of the captor. Inaction may be the only way to survive.

In the case of animals, there are additional factors to consider. Many people who live in countries where elephants are not indigenous see an elephant for the first time in the setting of a zoo or circus. Not only are these observers often ignorant of elephant ethology and natural history, but because captivity is culturally sanctioned, the human eye is psychologically conditioned to interpret captive elephant behavior as normal. Even if a person has seen elephants in their native savannahs on television or has visited them on "safari," the power of social conditioning can blind the viewer to what is going on inside the captive, and to the gradual degradation of an elephant mind and body. As a consequence, ceaseless swaying and pacing by captive elephants is labeled "boredom," but the same behavior observed in human captives is considered a psychological disorder disturbingly indicative of cruelly imposed stress.[5] What we are conditioned to see on the outside can tell a very different story from what is really going on inside a person.

These are the elephants, those likely candidates for complex PTSD, to whom Carol Buckley and Pat Derby decided to commit their lives. Carol started on her path toward becoming an elephant sanctuary director more than three decades ago, when she discovered Tarra. Tarra is now a full-grown adult living at the Sanctuary, but she was born in Burma and captured in 1974 at the tender age of six months.[6] She was flown to the United States and sold to a southern California tire dealer. Not long afterward, she met Carol, who was to become her champion and lifelong companion. Carol began caring for Tarra and training her, eventually purchasing her. For the next twenty years, the two traveled internationally, performing in various venues, including movies and television. Tarra's claim to fame was being the only elephant in the world to use roller skates.

Lucky for Tarra and other elephants, Carol had a change of heart and found her true calling. Buckley stopped working in the entertainment industry, and with her partner Blais bought the first parcel of the land that was to be the largest elephant sanctuary in the United States. In just over ten years, it grew to become a state-of-the-art facility by 2009, with room for a hundred elephants.

A steady stream of refugees has made its way to Tennessee. All have made startling transformations. A glance at the lush grounds of the Sanctuary reveals the obvious: there are abundant trees to wrestle to the ground, hills to climb and explore, brimming ponds for trunk baths and water play, a stable, permanent social group, and an array of cuisine to entice and nurture the discerning elephant palate. By all accounts, the Sanctuary is special. Amy Mayers, a media consultant from Washington, D.C., who devotes her spare time to the elephant cause, reflects on her first visit to Tennessee. "It was like walking on holy ground. There is a sense of the sacred."[7]

The Sanctuary has been home for twenty-four elephants, all females except for Ned, who arrived in 2009.[8] Like residents at a human retirement center, the elephants meet each other, often for the first time, having lived most of their lives in different communities. Nearly all are in the autumn or winter of life, like Winkie.

Winkie is an Asian elephant, born in Burma in 1966, small in stature, dark in complexion, who was retired from the Henry Vilas Zoo in Madison, Wisconsin, where she had lived for thirty-five years in a quarter-acre enclosure, surrounded by a moat, furnished only with two logs chained at the side of a shallow wading pond. She shared this area and a barn "the size of a two-car garage," according to the elephant advocate Lisa Kane, with an African elephant named Penny. Kane, a former Wisconsin state's attorney who spearheaded the successful campaign to liberate Winkie to Tennessee, says that the two were "chained in place for 16 hours out of every 24, and depending on the severity of the winter, they were confined to their cement barn for months on end." Coming to the Sanctuary, she has lost her companion Penny, but now lives with other "sisters."[9] Other elephants have been fortunate enough to find sanctuary *and* reunite with old friends.

Shirley and Jenny (not the Jenny from Dallas) first met two decades before their reunion at the Sanctuary. At that time, Jenny was a calf and Shirley a young woman in her twenties. Jenny was an Asian elephant, born and caught in the jungles of Sumatra and shipped for circus life at just over a year old. For the next ten years, she worked in the circus. Finally, the owners, exasperated by her multiple attempts to run away, lent her to the Hawthorn Corporation for breeding, to secure a new calf for the circus.[10]

Jenny's left hind leg was severely injured by the mating bull; left untreated, the injury became debilitating. Breeding was unsuccessful, and after a year she was sold to a small traveling circus. The injury to her leg worsened, and her health declined to the point where the circus owner had difficulty loading her in and out of the trailer. Eventually, she was just left inside, chained. Two years later, in April 1995, Jenny was found abandoned at an animal shelter in Las Vegas, dangerously underweight and with chronic foot rot. She remained chained for more than a year before the Sanctuary was made aware of her plight. At the time, however, the Sanctuary was in its infancy and needed to raise funds to be able to accept Jenny. Immediately, an emergency "Rescue Jenny Fund" was started. Although not all the necessary money was raised, the Sanctuary nonetheless rushed to take her in because of Jenny's precarious health. On September 11, 1996, Jenny arrived at her new lifetime home in the Tennessee hills, where Tarra's welcoming trunk greeted her.

Meanwhile, Shirley was living at Louisiana Purchase Gardens and Zoo in Monroe, Louisiana. Shirley, who was born in 1948, also was caught in Sumatra, but at the age of five years. For the next three decades, Shirley performed with the Carson and Barnes Circus, traveling the world and collecting adventures and tales of the dreary hardships of circus life. When she was just ten years old, the circus was held by Fidel Castro's soldiers in Cuba, and a few years later, while docked in Nova Scotia Bay, the troupe's boat caught fire, killing two circus animals; Shirley narrowly escaped but not before losing a large section of her right ear and sustaining other scars.

Like Jenny, Shirley suffered a major injury to a rear leg (her right, not the left), in her case when another elephant attacked her. Buckley points out that this assault occurred under severe duress, something that is often glossed over. The elephant who caused Shirley's injuries did so in a frantic attempt to reach an elephant friend whom circus personnel had moved away. This story provides another example of how the terms *elephant rage* or *violence* fail to capture what underlies and motivates elephant behavior: in this case, the desperate response of someone in fear of losing a loved one. The intensity of the attacking elephant's actions make even more sense when we consider the role of relationships in captivity. The majority of concentration camp survivors had formed close relationships; the unit of survival, therefore, was less the individual than it was a pair.[11] To see one's friend taken away causes grief, but it also constitutes a threat of the gravest kind: to survival itself.

After Shirley was exhibited for a year with the circus's freak show, her worsened condition was deemed a liability. She was sold to the Louisiana zoo, where she spent the next two decades, largely in solitary quarters to protect her from further injury by the other elephants. Zoo personnel finally decided that Shirley needed to live at the Elephant Sanctuary, where she could receive the care and carefully mediated resocialization that would revitalize her health and happiness. Shortly thereafter, at fifty-one, Shirley moved to Hohenwald, Tennessee, where she is now the oldest resident, sixty-one in 2009.

Shirley first showed wariness when she arrived. Poised at the door of the truck, she looked around and tested the air with her trunk; it was nearly two hours before she finally walked down the ramp to the barn, accompanied by her erstwhile zookeeper Solomon James. Elephants who are former circus performers are used to traveling from city to city, and accustomed to diverse living conditions. Zoo elephants generally have a less hectic life, but they, too, know life on the road, as they are occasionally shuffled between zoos to even out elephant numbers at exhibits, to accommodate a breeding program, or, like Jenny from Dallas, to dispose of what is considered to be a "problem."

But while life in the circus or zoo has taught wariness—to view change with suspicion—what elephants may feel when coming to the Sanctuary is closer to weariness. Elie Wiesel describes the evolution of such weariness and the focused numbness that creeps up as time goes by under the grinding violence of captivity and the past life begins to recede into an unfamiliar reality:

> The absent no longer entered our thoughts. One spoke of them—who knows what happened to them?—but their fate was not on our minds. We were incapable of thinking. Our senses were numbed, everything was fading into a fog. We no longer clung to anything. The instincts of self-preservation, of self-defense, of pride, had all deserted us. In one terrifying moment of lucidity, I thought of us as damned souls wandering through the void, souls condemned to wander through space until the end of time, seeking redemption, seeking oblivion, without any hope of finding either.[12]

Shirley finally settled into the comfy elephant barn with watermelon and other elephant-tempting treats, and was given a cool shower after the long trip. Then she rested, and Tarra went in to meet her. They intertwined trunks and Shirley guided Tarra to every injury she had sustained. According to Carol, the two "purred" as Tarra gently touched scars on the aging body. Then it happened.

At about seven in the evening, Jenny came into the barn. It does not do to try and render the story in a voice other than that of someone who witnessed the reunion and who knows elephants intimately. This is how Carol Buckley describes what she saw, the joining of two old friends after twenty-two years:

> There was an immediate urgency in Jenny's behavior. She wanted to get close to Shirley who was divided by two stalls. Once Shirley was allowed into the adjacent stall the interaction between her and Jenny became quite intense. Jenny wanted to get into the stall with Shirley desperately. She became agitated, banging on the gate and trying to climb through and over. After several minutes of touching and exploring each other, Shirley started to ROAR and I mean ROAR—Jenny joined in immediately. The interaction was dramatic, to say the least, with both elephants trying to climb in with each other and frantically touching each other through the bars. I have never experienced anything even close to this depth of emotion. We opened the gate and let them in together. . . . They are as one bonded physically together. One moves, and the other moves in unison. It is a miracle and joy to behold. All day yesterday (July 7) they moved side by side and when Jenny lay down, Shirley straddled her in the most obvious protective manner and shaded her body from the sun and harm. This relationship is intense and resembles that of mother and daughter. We are so blessed.[13]

Jenny died only a few years later, succumbing in 2006 to the injuries and illnesses she accumulated during her life with the circus. But there was joy for both elephants in their time together. When Jenny died, her loving friend Shirley was close by.

The Sanctuary is a place for dramatic change. The elephants come in one state and remarkably transform into another. One veterinarian remarked, "Carol gets the worst of the worst—the elephants whom the zoo fears will die on their hands or those so messed up mentally they can't be exhibited publically—those get sent to Carol." Yet these "worst-of-the-worsts" turn into the best-of-the bests. Elephants feared for their aggressive and dangerous ways with trainers become sweet-natured and affectionate; sullen, asocial elephants overnight seem to become the Sanctuary clowns, splashing and squirting water with their trunks and cavorting in the hills of their home. Release from fear into security presents an opportunity for an elephant's true nature to blossom. But does apparent psychological reversal mean that these elephants have recovered? *Recovery,* as discussed earlier, is a relative term that can be judged only from an individual perspective. Sanctuary elephants are a combination of whothey were, what they suffered, and who they were never able to become before. Their rehabilitation must be tailored with this understanding.

Elephant rehabilitation focuses on two main issues: the effects of forced physical confinement and the concomitant distortion of core relational processes and structures to which the prisoner is subjected. Physical scars may be soothed and softened, but as Jenny's premature death suggests, the severity of her wounds and illness could not be erased. As the trauma psychiatrist Bessel van der Kolk writes, what the mind cannot hold, the body does: "the body keeps the score."[14] What we see on the outside of an elephant hints at the map of suffering within, and like humans, every elephant has her own path and presentations in posttrauma life. Health issues derive directly from the harsh physical living conditions and indirectly from chronic psychological and emotional stress that undermines the immune system and the ability to heal. While physical needs are seen to, social rehabilitation also takes place. The challenges entailed are perhaps no better illustrated than in the case of the eight Sanctuary Divas, Queenie, Ronnie, Debbie, Minnie, Liz, Lottie, Frieda, and Billie.

In early 2006 routine at the Sanctuary was jostled by the simultaneous arrival of eight female elephants. In typical rescues, elephants arrive one by one, but this group came in four pairs. They had been trucked

650 miles after being seized from John Cuneo of the Hawthorn Corporation at the end of a contentious two-years' dispute with the U.S. Department of Agriculture.[15] Finally, Cuneo was cited for mistreatment and inadequate care in violation of the Animal Welfare Act.

At the Hawthorn Corporation barn the Divas had been kept in two groups: those labeled as dangerous, who had attacked or killed people in the past, were kept in the small stalls, Frieda's so small that she could not turn around. The others were kept in a separate room in chains. The eight females became a group for the first time at the Sanctuary. Social integration for other newcomers typically involves a single elephant joining an existing group, with an established set of social rules, after an appropriate adjustment period. The Divas, who had belonged to no previous groups, were starting from scratch. They were almost like strangers to one another, since they had not lived in a related group as free-ranging elephants do. Hence there was no set of collective rules to guide them. Older elephants might have had a taste of family life before capture, but there was never a secure loving bond to buffer their early shock. They went from losing mother and home to years of unrelenting hardship in a foreign culture. The Sanctuary provided their first opportunity to find out who they were, individually as well as collectively. The Divas have had to develop their own version of elephant culture without the accrued benefits of a free-ranging-elephant education.

During the Divas' first months in the Sanctuary, Buckley was concerned to see legacies of circus culture in the elephants' behavior. It is no wonder that after living for decades immersed in human behavior and values, elephants in captivity can take on human habits and psychology. We have seen in the example of the cross-fostering of Billy Jo the chimpanzee how readily a young mind absorbs the patterns and values of the culture in which it is immersed.

Carol discovered that Minnie was a "slapper"—an elephant who uses her trunk or body to "slap" circus personnel considered by their human colleagues as "low on the totem pole," typically men and women who feed and clean up after the animals.[16] While elephants only rarely dare to strike out at trainers, they have little compunction when it comes to the grooms because they know that there will be no reprisal. In fact, Buckley says, trainers consider it a joke when it happens. The object of abuse from trainers, elephants such as Minnie adopt behaviors of domination and take on values of the abusive culture in which they are forced to live, much like concentration camp kapos, prisoners who chose or were chosen to be guards over their fellow inmates as a way to survive.

Minnie continued to display violent behavior even after coming to the Sanctuary, where her abuse was also directed at her fellow elephants. Buckley recalls, "Minnie hit Debbie across the head with her trunk. Minnie slams into Debbie, one time knocking her feet out from under her and grabs Debbie's tail, ears, and sometimes trunk. It appears that Minnie is trying to inflict pain in the same body parts that I have observed trainers inflicting pain in elephants."[17]

Minnie's personality and behavior were very un-elephant-like compared with free-ranging individuals. The contrast should not be taken lightly. The abnormal behaviors and psychological states that incoming Sanctuary elephants express must be seen not as pathologies but as life narratives that communicate the bewilderment, agony, and struggle to survive after encountering the unspeakable. As the Holocaust survivor Viktor Frankl commented, "When we are no longer able to change a situation . . . we are challenged to change ourselves."[18] The statistically low number of acts of rebellion by elephants in captivity, including the rare human fatalities, is a product not only of elephants' well-known benevolent nature but also of the "terror, intermittent reward, isolation, and enforced dependency . . . [that] creat[e] a submissive and compliant prisoner."[19] Minnie and the other Divas were subjected to actual or perceived life-threatening brutality (a criterion of PTSD diagnosis), were fed only when staff members chose, were left chained alone or were otherwise separated from other elephants, unable to touch, and depended entirely on circus personnel for every aspect of their lives. Their compliance was exacted using methods employed by captors with human hostages.

But other experiences can undermine the soul even more. "The final step in the psychological control of the victim is not completed until she has been forced to violate her own moral principles and betray her basic human attachments. Psychologically, this is the most destructive of all coercive techniques, for the victim who has succumbed loathes herself. It is at this point, when the victim under duress participates in the sacrifice of others, that she is really 'broken.'"[20] There are occasions when humans do force one elephant to harm another, such as during *mela shikar,* the chase and capture of wild elephants by mahouts astride domesticated elephants (referred to as *kunkis* or *koonkis*); in some recent cases, kunkis have been used to drive away wild elephants as a method to resolve human-elephant conflict.[21] It is not known whether Minnie was ever coerced into hurting another elephant in the circus. Nonetheless, there is reason to believe that she experienced the equivalent of moral or self-betrayal when we consider her symptoms and explore hypotheses that might explain her behavior.

Minnie may have absorbed human culture to the extent that she identified with the trainers: she may have internalized the identity of a human abuser. Alternatively, she may have become a slapper as a way to gain a modicum of security by pleasing her trainers, or because slapping grooms was the only sanctioned means of getting back at humans in general and a way to release her "humiliated rage."[22] Her anger at the trainers who hurt her may have been redirected to and projected on the grooms who fed her. All of these scenarios are consistent with what has been learned from the testimonies of human captives.

It is significant, however, that when Minnie came to the Sanctuary, the environmental conditions that encouraged abnormal behavior disappeared—gone were threats of violence, chains, and abusive control—yet her behavior continued, expanding to include other elephants. Minnie began to show additional symptoms. Buckley, in her diaries, wrote: "Minnie appears depressed. It is as if she no longer feels any pleasure. Her face is blank. . . . She eats with the others but does not appear to be engaging with them. She shows no signs of pleasure. . . . Her play is solitary and frantic, and it always includes an object that she can stomp on and break."[23] Her continued abusive behavior toward Debbie and her dissociative, depressive periods both support the hypothesis that relates her behavior to an alteration in identity.

The difference between Minnie's personality and those of free-ranging elephants is not uncommon for one of her experiences. Human prisoners experiencing prolonged abuse often show "personality changes, including deformations of relatedness and identity."[24] William Niederland described concentration camp survivors whose "alterations of personal identity were a constant feature of the survivor syndrome. While the majority of . . . patients complained 'I am now a different person,' the most severely harmed stated simply 'I am not a person.'"[25] By adopting her captor's abusive behavior, Minnie, in effect, no longer *was:* no longer able to *be* an elephant.

During internment, the integrity of the self is shattered, "the values and ideals that lend a person a sense of coherence and purpose have been invaded and systematically broken down. . . . Even after release from captivity, the prisoner cannot assume her former identity. . . . The result, for most victims, is a contaminated identity."[26] Minnie's humanlike behavior directed at a conspecific, even to the point of inflicting pain in the same areas used by trainers, suggests that her circus identity was so ingrained that it remained tenacious even after she was physically removed from violent surroundings. Her profound depression, dissociation, apathy, and bursts of anger and rage correlate with symptoms of human prisoners who have experienced similar violations.

Was her violence against Debbie an expression of the hatred that many prisoners carry toward former captors? Do her solitary depressive periods indicate a profound alienation as a result of her betrayal of her elephant self? We cannot know exactly what either a human or an elephant is feeling. or what really motivates them. With humans, we have the advantage of verbal language, but as Elie Wiesel himself experienced, even that fails when there are things too incomprehensible to describe. In these surreal, desperate situations, the soul can become lost forever.[27] We do know, from Buckley and Blais's observations of other elephants at the Sanctuary who have come from similar circumstances, that Minnie experienced something at her core that remained intransigent after her move to Tennessee.

However, there are indications that Minnie has started to shed, or at least overcome, the identity she seems to have absorbed in the circus. Carol attributes one major shift to a defining moment, when the Diva Queenie suddenly died. Minnie and her friend Lottie were aggrieved and spent hours with Queenie's body after she had succumbed. Minnie

> painstakingly and gently lifted one front foot and then the other over Queenie's bloated belly and placed both feet centimeters from Queenie's back, arranging her hugeness protectively and completely over Queenie's body.
>
> Minnie breathed deeply, and as she exhaled . . . Minnie was letting go. There was a charge of energy in the air, and I would almost swear that in that moment Minnie had been transformed. Her love for Queenie allowed an opening in her heart through which all of her long-denied compassion, empathy and grief rushed in; she is now fully in the moment, the emotional moment, and she is standing confident with her dear friend facing her fears; she is not running away.[28]

Judith Herman recognized that most psychologists, from Pierre Janet in 1889 to the present, parse the progress of recovery into three main stages: safety, remembrance and mourning, and reconnection.[29] Minnie's ability to feel and express grief at her friend's passing, to "open her heart," is a sign of vulnerability, something that is considered part of the second stage of recovery made possible by the first, feeling safe.

Safety entails restoring a sense of control over one's mind and body and having a secure environment in which to live. The job of the Sanctuary is, therefore, to help Minnie and the other elephants achieve these goals. The barn, with the surrounding areas—the entire Sanctuary, in fact—is designed to provide the resident with the maximum sense of security and control over her environment. The Sanctuary protects its residents in one more, invisible, way: it prohibits open public access to the elephants.

At first, this strict rule may seem overprotective and exaggerated. After all, a gaze does not harm—does it? The elephants can roam at will all over the Sanctuary grounds, and people seeking to visit the elephants come in sympathy for their plight. But there is more to the rule than meets the eye, because modern humanity's insistent privilege for "right of sight" has been the foundation of captive industries and the cultural conventions of animal capture and display.

David Spurr refers to the "ideology of the gaze," a concept referred to in feminist literature as the objectifying "male gaze." Paraphrasing the French theorist André Malraux, he writes that colonial cultures' objectification is effected through the deliberate, commanding gaze. It is "a mode of thinking . . . wherein the world is radically transformed into an object of possession. . . . The gaze is never innocent or pure, never free of mediation by motives which may be judged noble or otherwise. . . . [It] is always in some sense colonizing the landscape, mastering and proportioning, fixing zones and poles, arranging and deepening the scene as the object of desire."[30]

The presumption of visual possession is the first step toward legitimizing physical possession; it thereby informs the rationale permitting objectification, and hence any fate that objectification can effect. Indeed, this is made clear with the common term used in zoos for where animals reside: *exhibits*. Rejecting visual privilege denies visual ownership and explicitly acknowledges elephant agency and its attendant right of privacy. The Sanctuary's founders are adamant that as soon as the elephants step onto Sanctuary grounds, they cease to be visual objects of human curiosity in the way that defined and condemned them to their former lives. The EleCam, live video footage of residents made available via the Sanctuary Web site, is part of the equipment used by the Sanctuary to monitor the residents' well-being, and although its images are available to the public eye, it offers only limited and remote viewing.

To help foster a sense of security, each stage of elephant rehabilitation is tailored to permit the individual to find her own pace and level of comfort. The resident can come and go when and where she pleases, and can socialize with whom she wishes. "The survivor's relationships with other people tend to oscillate between extremes as she attempts to establish a sense of safety": some days a resident might seek to surround herself with friends, other days she might prefer to be alone, a pattern of behavior that Minnie exhibited.[31] Sanctuary elephants are free to go in or out of the spacious and warm barns at will, or to seek shelter in the woods if they choose. The key word here is *choose*. The ability of the elephants to make their own choices helps cultivate the therapeutic alliance, a trusting and respectful relationship between the client and her therapist and part of recovery's process to reconnect.

Buckley describes the relationship between staff and elephant residents as "passive control"; its three key elements are space, time, and nondominating relationships. Sanctuary grounds are expansive, diverse, and healthful, and schedules are attuned to "elephant time," thus shifting the dynamic from a human-controlled routine to one largely determined by the pachyderm residents.

Elephants in zoos and circuses are told what to do in accordance with human schedules and are trained to raise their feet on command for veterinary checks. According to industry conventions, natural elephant relationships are based on pain and force, and to be trained, an elephant must be dominated from the start—a stern relational style that supposedly emulates elephant society in the wild. Buckley rejects that premise, and the elephant researcher Joyce Poole concurs: "I have no idea how this myth was started, but I have never seen calves 'disciplined.' . . . If a younger elephant, or in fact anyone in the family, has wronged another in some way, much comment and discussion follows. Sounds of the wronged individual being comforted are mixed with voices of reconciliation."[32]

Sanctuary staff members go to great pains to ask for, not insist on, cooperation. The process can take much longer than in a zoo, and may tax the caregivers to find creative ways to appeal to the animals' preferences, but eventually the approach is nearly always successful, and diverse health checks and applications are accomplished in mutual agreement. Buckley insists that no one on her staff is ever allowed to

> give an elephant a command or tell them what to do. What they can do is ask. We emphasize to the elephants that they have total freedom of choice. We will always try and find a way

> that will help an elephant decide to do something that they might need, such as medical treatment and foot soaks. So we put a bucket in the middle stall with all doors open. And bring goodies to attract the elephant over to the bucket and ask, "Will you please lift your foot? Will you lift up?" And it's critical that the caregiver believes that elephants possess the ability to make decisions.[33]

When maimed, infected, and cracked feet need to be bathed in soothing herbal waters, the recuperating resident is coaxed into the barn or to a locale outside and asked for consent to receive the care. The elephants are never pushed into doing something they are not prepared for.

When Billie, another of the Divas rescued from the Hawthorn Corporation, arrived at the Sanctuary, she was extremely frightened and nervous.[34] Born in 1953, she is one of the oldest elephants at the Sanctuary, but her long tenure in captivity had taken a psychic toll. At first she would not leave the barn and would jump at the sound of a vehicle engine or of a gate opening or closing. She would not permit staff to touch her, and so they did not. For fragile individuals like Billie, recovery can begin only when she becomes convinced that no punishment will be forthcoming and that she has control over herself and the environment.

Sanctuary staff is also trained to listen to the elephants so that they feel heard. The captive has been silenced as a hostage; being able to voice and assert herself is integral to dis-identifying with her victimhood and taking a step toward recovery. Part of being heard entails having one's needs and values met: receiving healthy foods, safe housing, and opportunities to form deep, lasting relationships. Elephants who come to the Sanctuary and PAWS never have to worry about losing their friends, until death do them part. Being heard is also reflected in the way staff members behave toward the elephants.

Buckley schools her staff so that "the elephants know that we are there listening, seeing, and responsive. For example, we are there when Barbara wants to drink out of the hose. It's her right to choose not to drink out of the trough. We are their servants. People in the [elephant] industry call it 'spoiling' and [say that] banging on the water trough is not acceptable. But we celebrate when someone bangs on the trough. They *should* be allowed to demand."[35]

However, the Sanctuary's "passive control" must not be interpreted as doing nothing. It is not sufficient for the therapist to be neutral: a lack of felt presence can create a void for the survivor and heighten her sense of disconnection that could further destabilize her psychologically. The survivor challenges the therapist to share in the struggle with "immense philosophical questions" raised posttrauma, and it is the role of the therapist to engage, "not to provide ready-made answers . . . but rather affirm a position of moral solidarity with the survivor." Such questions of philosophy involve helping the survivor understand the "rupture [effected by captivity] in her sense of belonging within a shared system of beliefs."[36] The therapist is participatory, a partner in dialogue who aids in the survivor's process of re-creating her self and negotiating past and present. For Minnie and the other Divas, the challenge is perhaps even greater because there was no shared system of beliefs other than inherited proclivities and the circus culture in which they matured and functioned: no stable structure to aid in making meaning or to guide collective rules of behavior.

Flora, another Sanctuary resident, provides a second, gentler but nonetheless sobering example of how deeply captivity affects elephants, and all animals. Flora is one of the youngest Sanctuary residents and one of only two African elephants living in the Tennessee hills.[37] She was born in 1982 and orphaned at two years of age during a cull in Zimbabwe. By the time she was three years old, she was a child star: she stepped out of the crate she arrived in and directly into a training ring and grew up to be what Carol calls "a daddy's girl." Ivor David Balding, founder of Circus Flora, toured Flora until he retired her to the Sanctuary when she was twenty-two years old. Her early retirement is testimony of Balding's care for her and his commitment to provide a better environment and future than he could offer.

Flora had a split personality—sweet, compliant, and affectionate when in the presence of her attachment figure, David, and aggressive and raging with elephants and humans alike when she was away from her "dad," as she was every year when she was sent to winter quarters. As she entered her teenage years, personnel at the camp and zoo (where she lived for more than two years before coming to the Sanctuary) began to comment on her angry outbursts. She was chained and given a wide berth. She also lived mostly among Asian, not other African, elephants. This is no small point because the two elephant cultures are as distinct as Italians and the English with the potential for conflict that corresponding differences in language, culture, and background can make. The cultural difference may also have contributed to misinterpretation and exaggeration of Flora's acting out.

African elephants have a reputation for being difficult and taciturn, but as Carol Buckley puts it, "They really aren't, they just do things and look at life differently." Where an Asian elephant will neatly push a bucket aside after drinking from it, an African elephant might crush it with one swipe of the front foot—not from any sense of pique, just as a matter of form, like one person crushing an empty can of Coke while someone else lobs her empty container intact into a recycling bin.

After being left at the Sanctuary and watching her former caregiver fade away in the distance yet again, Flora was beside herself. She screamed, ranted, and tore at her specially prepared enclosure, bereft. For many months, she could not be consoled or appeased. It has taken several years, with focused study for the Sanctuary to help Flora acculturate, and she has finally acquired a sense of proprietorship and behaves like the Sanctuary is her home.

Flora is one of the fortunate few who, while dominated and controlled, had a human owner who provided a sense of love, as evidenced by her obvious adoration of Dave and her "good girl" behavior. According to Buckley, Flora's trainer would "hook her" with an ankus in training, but unlike most other elephant trainers, he induced Flora's compliance with minimal pain and force, because her regard for him and desire to please were incentive enough: "Flora adored her care taker."[38]

Traumatologists working with former prisoners emphasize that captor-captive relationships are psychologically corrosive and volatile because of the intrinsic power differential. The total dependence of the prisoner on the captor can make the captor an omnipotent, larger-than-life savior in the prisoner's eyes. Psychiatrists describe this sort of relationship as "traumatic bonding."[39] As a performing elephant, Flora was controlled at all times. When and what she ate, where she went, what she did at every minute of the day and night were decided by someone else. Almost by definition, her mental state and abilities to function had to be entirely dependent on her trainer, because her very existence was determined by him. The isolation from other elephants or reduced level of interactions with conspecifics heightens the intensity of the bond. For Flora, Balding probably became the savior who "rescued" her every spring from the winter camp and took her home.

Cultivation of the keeper-elephant bond is considered vital for successful management, and many elephant keepers and trainers speak of their love for their charges; the lengths that people like David Balding go to better captive life provide testimony of their concern. But Fred Kurt, an Asian elephant ethologist who has worked with elephants in captivity in diverse settings for decades, is unconvinced. He says that claims of love and "lifelong bonds" between elephant and man belong "to the kingdom of beautiful legends."[40] Human-elephant bonds forged in captivity cannot be viewed in the same way as those outside forced confinement, when both parties are free agents. Captive relationships are contingent on the psychological and physical fetters of zoo and circus life.[41]

Flora also lived in social and emotional ambivalence, since the person with whom she identified was a source of both pain/rejection and care/acceptance. At one level, she was provided with secure love, but at another, this sense of security was an illusion because any day her world could collapse, and it did—every winter when she was left at winter camp. When such an individual grows up, the adolescent or adult may maintain a favorable image of the caregiver, as did Flora, but as John Bowlby observed, "at a less conscious level [the child] nurses a contrasting image in which his parent is represented as neglectful, or rejecting, or as ill-treating him."[42]

Consequently, instead of developing a secure attachment that provides a sense of self-confidence and an ability to empathize and interact well with others absent the presence of the attachment figure, Flora displayed behaviors and psychological states consistent with an insecure attachment. An insecure attachment develops when a child fails to have an internal representation of a responsive caregiver in times of need—when, for example, the child feels frightened or lacks needed reassurance, as Flora did in winter camp. Her trainer was unavailable—gone—and her world of love and joy transformed into a horrifying void occupied by threatening strangers. Flora exhibited characteristic symptoms: an inability to self-regulate, irritable outbursts, an overdependent and intense relationship, difficulty forming other relationships, a difficult personality, volatility, and slow adaptability to change.[43]

It might be appropriate to view Flora's outward-directed frustrations as being childlike. In her role as a performing elephant, she had failed, even past the age of twenty, to individuate sufficiently to contain and regulate her emotions, and to achieve a sense of self distinct from that of a submissive appendage to her trainer. Physically and emotionally, she was caught in the amber of perpetual childhood through her role as a performing animal subject to someone else's agenda.

In her angry, desperate outbursts she voiced fear like a child terrified by loss, leading to a profound feeling of impotence. However, her resultant rages must not be reduced to temper tantrums. Often, animals in captivity are belittled through infantilization, their screaming, aggression, and other displays dismissed as "baby tantrums." Buckley has seen in the cases of even the far-from-baby-sized elephants that "there is a tendency to infantilize and control protectively as if they are not fully functioning adults. So the [Sanctuary] staff is told never to talk baby-talk to elephants. They can be touched and cuddled, but they are not to be treated as a baby."[44] The simplest and most direct way to gain an understanding of what might lie behind such displays psychologically and emotionally is to put ourselves in the animals' cages—looking out at the determinist and uncontrollable world beyond the bars. A captive individual has little recourse to express herself and limited ways to cope with helplessness other than rage, dissociation, stereotypy, withdrawal, or, as Jenny from Dallas exhibits, self-injury.

We can also understand the evolution of Flora's behavior from the perspective of neuropsychology. Stephen Porges developed what he named the "polyvagal theory" to describe the complexity of the autonomic nervous system that is engaged in regulating arousal states.[45] The autonomic system, particularly that of altricial species—those whose young strongly rely on early social interactions—is more than a simple "on-off" switch. Instead, it is governed by a hierarchy of subsystems: the sympathetic systems (mobilization of resource: "flight or fight"), the dorsal parasympathetic system (immobilization), and the ventral parasympathetic system (social engagement). This last subsystem provides an interactive way to engage (or disengage) with the environment without taxing the other two systems, thereby providing flexibility and reactive continuity, as in the case of Dame Daphne's orphans, who would be considered as having a well-functioning social engagement system. When there is trauma, particularly in the absence of a restorative secure attachment, this subsystem is poorly "tuned" and overridden, resulting in states of hyper- or hypoarousal under stress, either of which strains an individual's physical and mental health, as well as creating dissonance with the environment. This scenario and neuropsychology fit what we know of Flora's upbringing and behavior.[46]

Flora's upbringing deviated from normative elephant rearing in another way: the social context she experienced was not that of free-ranging elephant culture. She had a single attachment figure, her owner, as opposed to the constellation of allomothers that wild infants experience. In a sense, what Flora inherited (elephant nature) was mismatched with what she experienced in her rearing (human captive nurture), much like the experience of Billy Jo, but again with the significant difference that she did not have to experience the severe trauma that the chimpanzee sustained after he was given up to a laboratory.

The Sanctuary faced the formidable task of facilitating the healthful development of a nearly grown female elephant. Sandra de Rek, the Sanctuary's African elephant keeper, became Flora's private psychotherapist. Sandra had worked in Europe with elephants in captivity before coming to Tennessee. For more than three years, she spent day in day out with Flora, talking with her, listening to her, and tending to the innumerable details involved in an elephant's care. Flora's reputation and volatility had given Sanctuary staff what Carol Buckley calls "healthy fear." For instance, Flora would look calm and content while engaged in eating, then suddenly become enraged, focusing her wrath on any unlucky person who happened to be around. At times, she even would bang her head repeatedly on a steel gate in anger or frustration.

Flora remained edgy and mistrustful of Sandra for a long time, and circumspect about any new experiences or changes in her environment. For example, after Flora had seemed to adjust to the barn area, the Sanctuary staff opened the gates to let her begin exploration of the adjacent fields. Most residents—soon, if not immediately—become enthralled with their newfound freedom and set out to explore the open world of the Sanctuary acreage. Flora, on the other hand, became nervous and proceeded to methodically—"obsessively," according to Carol and Sandra—dismantle the gate and fence bolt by bolt. Pieces of fence were not carelessly scattered but were neatly piled. Much to the disappointment of weary staff, Flora did this repeatedly, thereby requiring a rebuilding of gates and fencing each time, while she was returned to the confines of the small yard and barn. This was puzzling to both Buckley and de Rek, and contrary to their experiences. How could an elephant prefer confined space? When Flora was in the small yard, she was happy and not fearful, and her play was not destructive.

If, in the mind's eye, Flora's previous environment is superimposed over the present, we can get an idea of what might underlie her obsessive behavior. The new sanctuary world lacked the bounding support that her intimate bond with her former owner and circumscribed circus life had provided. Life there might have been controlling, but it was familiar and well defined, and therefore it felt safe. Then suddenly, when her attachment figure vanished, she was left feeling unbonded and unbounded. Her angry outbursts, her reluctance to form a relationship with anyone, and her obvious panic at the prospect of an ill-defined environment suggests a sense of a lack of control, extreme vulnerability, and disorientation. Flora lost her points of reference, both external (Balding) and internal (an identity referent to Balding). Picking apart the gate and obsessively stacking its pieces might reflect her effort to gain control, to shut out the fearful abyss that open freedom may have presented. By focusing on the gate she could perhaps, as Sylvia Plath wrote, "shut [her] eyes" so that "all the world drops dead"; when she finished dismantling the gate, she might hope to "lift [her] lids" so that "all is born again"—return to life with her former owner.[47]

Day by day, Sandra worked with Flora, and the young elephant no longer felt the need to retreat to her confined area when challenged. Sandra provided a new attachment figure with whom Flora could interact, but one who was egalitarian. Gradually, through what Sandra refers to as tests and trials, Flora began to trust her caregiver and explore their relationship and her emerging sense of self. Eventually, Flora did happily venture out to the spacious open area, and today she explores, plays, knocks down trees, and socializes with the other African resident, Tange. Flora has begun to mature and come into her own.[48] Like the Sheldrick orphans, Flora is lucky because she has years ahead of her that will permit a ripening of her self and identity, and the decades anticipated in sanctuary promise to overshadow any earlier bumps on her road.

Much has been learned and much more remains to be learned about and from the elephants in sanctuary. We have seen that it is legitimate and informative to perform a clinical diagnosis on elephants. Hypothesizing a diagnosis of complex PTSD for the Sanctuary elephants is useful because it incorporates recognition of what they have endured and acknowledgment of the importance (and first, the existence) of subjective elephant experience. By mechanistically linking relationships between past and present, traumatology provides insights into behavior that otherwise would be vilified, its perpetrator summarily eliminated, as has been done so many times with elephants in captivity like Black Diamond. Winkie is one elephant who has profited from psychology.

She killed a much loved caregiver, Joanna Burke. Winkie had never exhibited dangerous behavior before. Given what we know of Winkie's past and of traumatology, she probably experienced what is called a PTSD "flashback," wherein a survivor is catapulted back to the reality of her trauma as a result of some trigger—a smell, a sound, or, in Winkie's case, a possible distortion of sight as the result of a minor eye injury. In this distorted frame of mind, the survivor reacts to what she thinks is life-threatening reality, unaware it is a memory. Winkie suddenly lashed out at her caregiver. For weeks afterward, Winkie was lost in profound depression.[49]

Viewing animals through the science of the mind will also bring challenges as well as insights. The two make a powerful mix, and the debates concerning PTSD, and psychiatry in general, are likely to spill into the new field of trans-species science. We must also be cautioned by one of the lessons that exploring elephant psychology and transcultural psychology has taught: individuals who come from diverse backgrounds can be understood as equals, but their differences must not be lost. If formal diagnoses are to be followed as laid out in the DSM-TR, then transcultural and, now, trans-species psychiatry demand an expansion of that system to accommodate the range of meanings and presentations that trauma can create. Cathy Caruth, who writes extensively about trauma, speaks to the ambivalent knowledge that traumatology has introduced. PTSD and trauma in general have provided

> a category of diagnosis so powerful that it has seemed to engulf everything around it: suddenly responses not only to combat and natural catastrophes but also to rape, child abuse, and a number of other violent occurrences have been understood in terms of PTSD. . . . On the other hand, this powerful new tool has provided anything but a solid explanation of disease: indeed, the impact of trauma as a concept and a category, if it has helped diagnosis, has done so only at the cost of a fundamental disruption in our received modes of understanding and of cure, and a challenge to our very comprehension of what constitutes pathology.[50]

This stance implies that psychiatric diagnoses are best used as hypotheses, and we would do well to take this philosophical view when approaching therapeutic design and understanding elephant symptomology.

Delhi, who was born in India in 1946 and died peacefully at the Sanctuary at age sixty-two, provides one sterling example. She unwittingly became the focus of a landmark case in elephant rights—the first elephant in captivity to be seized from her owners for inhumane treatment and years of abuse.[51] Her suffering was so egregious that the USDA performed an emergency intervention.

How are we to diagnose Delhi? Technically, she is an outstanding candidate for a diagnosis of complex PTSD, with a range of symptoms and so-called disorders. In some ways, however, the question of discrete diagnosis becomes moot. Instead, we should look at the person Delhi has been, a person who went through horrific ordeals, survived, and emerged psychologically and emotionally the way the Sanctuary met her. A tribute left on the Sanctuary Web site illustrates this sentiment: "You spent 58 years in horrible bondage. You were beaten, treated cruelly, chained and suffered with diseased feet. Yet miraculously, you forgave the past injustices and once you came to [the Sanctuary], you lived each day to its fullest. You weren't neurotic or fearful, even though you had a million reasons to be suspicious and hateful towards humans. Your enormous heart and beautiful spirit allowed you to live in the present and enjoy every moment of the freedom granted you."[52] Here is where the true lessons lie, in the heart—lessons that teach us how to better care for one another, and critical lessons about how we can change so that Delhi's experiences never happen again to someone else.

10
Speaking in Tongues

As she approached his enclosure, her body stiffened and a cold sweat broke out. Massa would agitate and vocalize with his trunk swaying. It took every ounce of willpower to stop from just throwing the food and running. One day, Anna willed herself to stand still and kept eye contact with the approaching mountain. Suddenly, something happened. Anna began to cry, tears streaming down her cheeks and neck. The bull came to a standstill motionless in front of her and slowly reached out with his trunk. Anna reached through to touch it. Later, she recounted: "At that moment, it was as if the world melted. My fear disappeared and I opened my heart." Massa was still regarded as someone to watch with care, but he gradually began to relax, and acted almost eager to touch and be touched through the fence. He started playing and pushed the huge rubber ball over to Anna in fun.

When Ralph Waldo Emerson wrote that people see only what they are prepared to see, he was probably not imagining today's virtual world of talking screens, fiber optic relationships, and Blackberry scribes. The degree to which technology has allowed sight to eclipse all other senses can be measured by a simple excursion to the country. We venture into the woods, and we alight to stand in the still; there is a sense of emptiness. Gradually, senses adjust from the din of honking horns and caroling mobiles and tune to the wild. We begin to notice light refracting down through treetop stairways and the startling flash of red as a woodpecker dashes by, the faint hollow staccato of an owl, cold flecks of water sprayed from the creek, the mustiness of fallen leaves, and the taste of salt as tongue passes over lip. Skin prickles, heart quickens, eyes widen, we no longer feel apart, and we begin to merge with the watchful sensuality of feral life.

Moments like these are few for the urban dweller. Most of the time it is the mind, not the body, that dominates and must deal with the challenges of everyday survival. Similar to the human body, animals are kept in place by strict rules of behavior—elephants live in zoos, circus rings, and *Out of Africa,* dogs belong on a leash, parrots sit in cages, bears are seen on Forest Service posters and Yellowstone Park postcards, turkeys are framed on holiday platters, and so forth, species by species. Violations of these boundaries are eyed suspiciously if tolerated at all. Tyke, the breakaway circus elephant, was gunned down, unleashed dogs are impounded, biting and screaming parrots are taken back to the pet store, garbage-rummaging bears are shot, and wild turkeys soiling patios are summarily dispensed. Humans who stray across conventional boundaries are also penalized. Hunters get licenses to shoot deer, but those who feed deer are prosecuted or jailed; and the general opinion was that the mauling death of Timothy Treadwell, the Grizzly Man, by his beloved bears, was just deserts for wrongful intimacy with a wild animal. The line between human and animal exists not only in concept but also in body.

Consequently, when a human body and that of an animal intersect, there is a sense of unease. Whether it is because sensuousness is so often equated with sexuality, or because blurring the line between ourselves and other animals threatens our sense of humanity, intimacy carries a taint of the forbidden. It is within this liminal zone that Elke Riesterer treads in her healing work with elephants.

Riesterer is a massage therapist who uses the medium of bones and flesh to soothe human minds. In 1997 she expanded her work to embrace the world of the furred, feathered, and scaled. She makes a twice-monthly pilgrimage from her home in Santa Cruz, California, to tend to elephants, tortoises, monitor lizards, giraffes, snakes, and emus at the Oakland Zoo. Elke even travels to Kenya, Thailand, and India to dialogue with ailing Asian and African elephants using her "listening hands."[1]

One of the first images at the zoo to greet her eyes is the oval dome of O. J., an Aldabra tortoise who is more than one hundred years old and hails from the Aldabra Atolls, Madagascar, silhouetted against the sunlit concrete wall of his exhibit.[2] Second only to the Galapagos tortoises in size, the adult male Aldabra can weigh more than 500 pounds, with a shell spanning up to five feet in length. Female tortoises weigh up to 350 pounds and have shells nearing three feet. Their dark brownish black-to-gray shell protects the long neck that can extend almost three feet past the shell lip. In their native African habitat, using strong scaled front and back legs for leverage, Aldabra tortoises like O. J. knock down shrubs and small trees in search of juicy roots and scrambling insects nestled in the uprooted plants. O. J. lives with Ralph, his junior by a few years, and a newcomer, Gus. Next door are the three female "babies" of the family, T, B, and Gigi. T and B are only half O. J.'s age.

The feel of sliding python skin and wrinkled tortoise necks is not usually associated with pleasure. In fact, the sensation is liable to evoke just the opposite reaction for many people. Neither do these animals seem to present an inviting surface for massage. But Elke's eyes have learned how to see the body in ways that elude most of us. Under her healing hands even the most cold-blooded become pliant and warm.

When Elke first met T, the young tortoise was shy and withdrew her head into the shell. Now when she hears Elke's voice and senses her coming, T stretches her neck out. Sitting besides "Little T," as Elke affectionately calls her, the therapist moves her hands slowly around and over the shell. Continuing with the same fluid motion, her touch lightly dances down the curved front legs and up to T's mouth as Elke offers a slice of glistening melon. Gracefully arching her neck, T takes the fruit and methodically munches and swallows. The sweet juice runs down her beaked face, falling in tiny golden drops to the earth beneath her claws. After T has finished eating, Elke begins to gently move her hand up to the lined neck to find the most vulnerable part of a tortoise's body: the flesh at the joining of neck and her upper body. The skin here is warm and soft, in stark contrast with her protective shell and armored legs. T turns her snakelike head toward Elke and gazes at her with obsidian eyes. Gradually, T's legs drop a little and her eyes blink shut restfully as Elke strokes her neck in tiny circular motions.

Making these magical circles is one of the many techniques of TTouch, a method named for and developed by Linda Tellington-Jones to work with stressed and injured horses.[3] While TTouch is often called a type of massage, Riesterer distinguishes between massage, which has a muscular focus, and the seemingly insignificant motions of TTouch, which concentrate on the nervous system. TTouch, she says, is a type of energy work exercised at the body's surface.

Body-centered therapists often use multiple styles. In addition to TTouch, Elke draws from methods including polarity, acupressure, and Swedish massage to create a unique blend tailored to fit the form and needs of individual bodies. Entering a heated reptile enclosure, Elke visits with Garline, the monitor lizard, who is being fed tasty garden snails by the keeper. "After finishing the last of the snails," Elke says, "Garline dropped down from her rocky ledge and slowly and deliberately, walked over to another area within the exhibit."

> At first, she shied away from touch, but then began flicking her tongue over my hand and arm to see who I was. I spoke to her in a low tone and lightly touched the top of her head that has a specific spot known as the third eye and considered a

> critical point in the world of acupressure. She instantly relaxed and closed her eyes. I continued with a light massage along her spine to the base of her tail, gently TTouching. Then I massaged just under her arms and along the side of the belly down to knead the fleshy parts of her feet, alternating the TTouch circles all the way to her elongated nail tips. Garline shut her eyes and went completely limp.[4]

For first-time treatments and more timid clients, Elke begins by using feathers or, with oversized clients like elephants, soft tree branches as extensions of her arms. Hand-held branches drawn down and across the back and sides simulate the scratch and pulls of acacia as an African elephant pushes through savannah brush or an Asian elephant through lush foliage. Outside of places like PAWS and the Elephant Sanctuary in Tennessee, elephants in captivity are rarely afforded such sensory luxuries. Most zoos are amid towns and cities whose reverberating sights and sounds are nothing like those of native habitat. Indeed, it is somewhat unsettling to imagine what elephants' perceptive feet and ears may detect from the maze of pounding pavement, creaking buildings, and buzzing wires surrounding them. Even within the zoo's confines, there is little escape from human intrusion. For those of us accustomed to living encased in the cement, glass, and synthetics of urban life, zoo exhibits may look only somewhat stark, but they must be bone-penetratingly bare for their animal residents.

Our cultural emphasis on sight makes us forget that animals are born to live in a dense matrix of touch, smell, taste, and sound as well, their bodies interleaving with the environment. Who knows what wealth of information is exchanged between two elephants in passing—what they have each eaten, their state of health, and the log of travels and community news conveyed through passing molecules from trunk to tongue.

Ray Ryan, an elephant keeper for four and a half years at the San Diego Wild Animal Park during the late eighties, speaks of perceptive elephant tongues. Ryan grew up on the south side of Chicago, a street-savvy wise guy but at the same time lovingly caring for hamsters and a menagerie of rescued animals. He began his profession caring for primates at San Diego State University, where he completed his degree in psychology and watched over monkeys used in studies. Later he joined the staff at Pat Derby's PAWS. When he was hired at the Wild Animal Park, Ray thought he would continue as a primate keeper, but upon arrival he found out that that he had been given a different assignment: elephants. It was one of those simple twists of fate that would change his life.

Ryan's charge was a group of African elephants headed by the matriarch Peach, a canny elephant with a commanding personality who, with the rest of the elephants, was well versed in the ways of their human keepers. Ryan relates that when he and his boss sometimes came into work after a late night, the elephants knew what had been going on:

> They knew all right. When elephants meet they put their trunks into each other's mouths and transmit breath, energy, and thoughts through their tongues like the front part of a stereo woofer. They can read you too, just like someone from the south side when you look into each other's eyes. We would come in and like always, the eles would reach out their trunks to our toes, crotches, fingers, and armpits—where you sweat. What one ele found out was passed on to another and transmitted down the line. They wouldn't be mean, but they would let us know they knew how we felt and what we had been doing by walking us into the walls—you know, like saying "we know you are screwed up, but we've been chained up for over fourteen hours, and we're sick of it, so get going."[5]

In forced confinement, mind and body must renegotiate with the environment to survive the radical change from nature's own conditions. The hostage must somehow manage the panic that comes with being totally controlled. The body ceases to be an agent of self-determination, becoming instead a burdensome liability. It is currency in the transaction of routine and lives at the mercy of physical pain. Elie Wiesel speaks of the alienation and dissociation that can develop under such severe duress. "I was dragging this emaciated body that was still such a weight. If only I could have shed it! Though I tried to put it out of my mind, I couldn't help thinking that there were two of us: my body and I. And I hated that body."[6]

Carol Buckley says that many elephants, in their first days, sometimes weeks, at the Elephant Sanctuary, have faraway looks and appear detached. She explains that in the circus they have learned not to feel, or to care about other elephants, or to make attachments. "Anything that you care about makes you vulnerable. You become desensitized and anything new has a bad association."[7] Buckley's descriptions are eerily similar to the accounts of prisoners. "New" spells change, and most if not all change for a prisoner has been bad. Wiesel, early in his and his father's incarceration, grew hopeful when there was a change—perhaps they would be released, reunited with loved ones, moved to a better camp—but he learned that to quicken when something new stirred led only to more pain.

Touch can therefore be a daring move. The awakening body may provide the survivor with his first opportunity to identify with a self that is someone other than an inanimate object of violence; to reclaim authority as a living participant and be able to initiate and make choices is to refuse to be victimized. Yet reaching that turning point may involve giving up perceptions and behavior that make survival possible. For the captive, the caress of a hand or trunk represents revitalization and reconnection with body and emotions, but also vulnerability. To open oneself, to feel, means having something that can be taken away. There are other dangers. Healing touch, even the gentle ministrations of Elke's, can be confused with exploitive manipulation.

Artificial insemination is an increasingly common method used by captive institutions. As importation of wild-caught elephants became more difficult, zoos began to expand captive-breeding programs to keep up stock availability. Furthermore, while captivity appears to have less impact on male reproductive systems than on those of females, males have shown marked lack of libido and suboptimal sperm quality, in part, perhaps, because of social suppression under the domination of animal handlers.[8] Artificial insemination is used to overcome these constraints.

Osh is a young bull elephant who was brought from England to the Oakland Zoo. In compliance with AZA guidelines, he undergoes anal training to ready him for the sperm collection procedure. The elephant is manually stimulated until ejaculation, at which time his sperm is collected in a vial; later it is used to inseminate a female elephant via a long, endoscope-guided tube inserted into her vagina. Of the more than one hundred elephants on whom Riesterer has used body therapy, Osh alone has shown anxiety or discomfort when she tries to massage his tail and buttocks; he is also the only one of her clients who has undergone anal training. Riesterer describes his wariness as "an initial uncertainty expressed in jerky motions of not knowing what may come," an uneasiness that may reflect an association of her touch with the invasive anal procedure.[9] In Riesterer's experience, elephants are typically relaxed by massage in this area of the body.[10]

Emerging into freedom from a South African jail, a former prisoner related that after years of control, the mind conforms to the bars, walls, and guards: conditioned fear and shutting down had become so ingrained that physical restraint was almost unnecessary.[11] How then can Riesterer hope to have her work make a difference for the captive or to encourage well-being in the continuing presence of the structures and agents of trauma, captivity and captors?

From the point of view of traumatology, therapy conducted in zoos or Indian elephant labor camps cannot be expected to achieve recovery when recovery, by definition, demands the end of captivity itself.[12] In theory, encouraging the relaxation of psychological and emotional defenses might cause iterative retraumatization. On the other hand, despite theoretical misgivings about massage and other body-centered therapies, it is obvious from the response of long-term clients that Riesterer's work is welcomed. She reflects that "mindful therapeutic touch is a potent enrichment of their severely compromised lives and the calming effects are truly visible and palpable. I am inclined to name it a 'bandage of comfort,' a type of therapeutic triage in a place with no way to escape. It is perhaps much like the few progressive human prison facilities where the residents receive some form of therapy yet still must live and deal with the hardship of their imprisonment."[13] For some, healing touch is one of the few moments of social contact.

Soila is a teenage African spurred tortoise at the Oakland Zoo. One day, Soila's keeper asked Elke to take a look at the young tortoise. The keeper was concerned because Soila had stopped eating and defecating. Elke slowly approached Soila, sat down next to her, and began massaging her along the head meridian, down the back of her shell by using revitalizing friction (which "she absolutely loves and responds by wiggling into my hands"), down to her feet and the base of tail. After forty minutes, Soila suddenly looked up and walked over to the fruits and vegetables that had been provided, ate a few pieces of zucchini, and then left well-formed tortoise scat.

When an acquaintance, a specialist in child psychiatry, heard the story, she was reminded of several young patients who, when unable to feel control over their environment, controlled their bodies and showed similar symptoms. "Teenagers are bursting to explore the big wide world and come into their own. In a depauperate environment devoid of social contact, you just start shutting down."[14] For someone like Soila or Garline, the intimacy of touch is healing to heart and body. For other seasoned residents, Elke's ministrations offer something additional: the therapeutic touch may relax the client for a few minutes and allow her to recall pleasant memories of brothers, sisters, and mother leaning up close and caressing her face, mouth, and ears.

The healer can play a crucial role by redefining connection as a positive experience and offering a neutral, protected space where the "old self," which is still needed to endure captive survival, and a "new" post-trauma self may begin to coexist. While many facets of an elephant's relational self can never be replaced—the self with mother and self with family, the self in the world of origin—their functions might be. Caring touch may provide a chance, particularly in sanctuary, to stimulate the life force that has been suppressed, to create a buffer from the damaging environment, to give a reason to feel and seek pleasure again, and to be alive.

The story of Massa and Anna echoes the experiences of trauma therapists and their human clients everywhere. It is at defining moments of breakthrough like the one when Anna's heart "opened" that trust is established and the healing process can begin in earnest. The therapist-client bond forms the cornerstone of all healing. Without deep trust and a sense of security in the therapeutic alliance, recovery cannot proceed.

One of the most common media that fosters a bond is the human survivor's narrative. The Holocaust survivor and psychiatrist Dori Laub refers to it as the "imperative to tell," the need to relate the experience of survival as well as the impetus to tell it in order to survive.[15] Telling requires a listener to bring coherency, a sense of reality, meaning, and place outside the confines of trauma. C. G. Jung writes, "The mere rehearsal of the experience does not itself possess a curative effect: the experience must be rehearsed in the presence of the doctor."[16] But the survivor's narrative must not be envisioned too narrowly. Verbal expression is used most frequently in Western therapies and Western culture in general, but this tradition has been reshaped considerably by perspectives from other cultural practices and by research in neuroscience and psychology.

Neuroimaging has made possible more discoveries, bringing visual clarity to murky theoretical waters. Sigmund Freud's conceptualization of the mind as made up of the conscious (what is knowingly accessible), preconscious (what is accessible but not yet conscious), and unconscious (inaccessible) resonates with what has been uncovered via neuroscience.[17] In one modality, the analyst and analysand engage in dialogue at the conscious level, the content and stuff of everyday life. This is the realm and medium of science—the purposeful and aware states of mind and action in which individuals engage. But while the brain's left hemisphere may be largely in charge of the conscious, explicit, verbal world, the right hemisphere is busy conducting the same sort of business at different levels and modalities: the unconscious, implicit, emotional world.

Behind the exchange of symbols are messages from the unconscious: that part of the mind and body outside a certain awareness comprising, Freud maintained, repressed memories, thoughts, emotions, and knowledge beyond the personal self. Depth psychologists—those who make a study of the unconscious—refer to the second body of knowledge as the collective unconscious, which houses archetypal images and instincts, or what we might think of as our mental heritage. Experiences and emotions are embedded in the unconscious like sedimentary strata at a lake's bottom and their presence becomes known only when they surface into conscious awareness. The unconscious communicates in a number of ways, including dreams, active imagination, symbols, and—relevant to recovery work with elephants—somatic and psychological symptoms. Therapists like Riesterer tap into these embodied memories that exist beyond words: "The target of my work is not the ailing foot or aching back, nor even person or elephant, so much as it is that individual living deep inside, maybe it is the soul, who is in need of healing. We don't really have language today that meaningfully describes what goes on between a human being and an animal. So much is communicated and known through a glance, through warmth, energy, and the flex of a muscle that words and even concepts don't capture."[18]

Allan Schore, who has helped to galvanize the evolving field of neuropsychoanalysis, describes human-to-human exchanges in much the same way, using different language and images: "During the treatment, the empathic therapist is consciously, explicitly attending to the patient's verbalizations in order to objectively diagnose and rationalize the patient's dysregulating symptomatology. But she is also listening and interacting at another level, an experience near subjective level, one that implicitly processes moment-to-moment socioemotional information at levels beneath awareness."[19]

While verbal exchange flows consciously, unconscious dialogue is mediated by the right half of the brain, which picks up the emotional content of words and interprets what is being said in a broader sense. This information is that kind of niggling, nuanced feeling that arises when your spouse's ex tells you how great it is to meet you, or the taxi driver insists that the restaurant is only a block or two away via this little-known shortcut. Our right brain excels at detecting another's emotional state, which may or may not match the words being said, by observing imperceptible flicks of facial shadows and expressions, posture, and prosody;

the body feels this information by so-called somatic or kinesthetic experiences. As Michael Mathew writes, "The body is clearly an instrument of physical processes, an instrument that can hear, see, touch and smell the world around us. This sensitive instrument also has the ability to tune in to the psyche: to listen to its subtle voice, hear its silent music and search into its darkness for meaning."[20]

What we learned from John Bowlby's descriptions of infant-mother attachment about movement, facial expression, tone of voice, scent, rhythmic and emotional changes applies to the therapeutic setting. Right-to-right brain communications operate much like the intricate duet of mother and child, or the intersubjective exchange between survivor and therapist. Clinical and scientific studies converge on the same conclusion: "Intersubjectivity, a central construct of current developmental, clinical, and neuropsychoanalysis, is more than a match or communication of explicit cognitions, and . . . the intersubjective field co-constructed by two individuals includes not just two minds but two bodies; the body is the very basis of human subjectivity. In other words, at the psychobiological core of the intersubjective field is the attachment bond of emotional communication and interactive regulation."[21] This multimodal dialectic of brain-mind-body between two individuals relates to what Freud and Jung called "transference-countertransference."

Ray Ryan often describes talking with elephants and marvels at how much goes on in the realm of silence and movement, at levels and in ways that have no names: "The real communication happens just looking at each other, touching, smelling—and they're all connected. That's why when they arrive at Carol's [the Elephant Sanctuary in Tennessee] some already know each other because of the 'ele underground railroad.' One elephant meets another at the zoo, she's moved and then passes on the information and news to the new elephants she meets, and then they tell others and so on. So when two elephants meet, even if they haven't met before, they may have heard about each other from the others maybe years before."[22]

But what about other beings who appear to have different languages and ways of communicating? How does the therapist work with the trauma survivor when she is not human? Elke describes a session with an elephant in captivity:

> I like to say mouth work is a very sensual experience for an elephant as I have seen the same type of response in many elephants over the years. It can be a profound way by acting as an entryway into a Body Therapy session and open an individual animal to a deeper mental and emotional experience. It often acts as the fastest way to introduce a trance state with certain animals. May be that it reminds them of being young sucking at their mother or how they once were able to greet each other in the herd in the traditional elephant way of touching trunk to mouth. I use an orange piece while at the same time massaging fingers moving gently up and down her rosy, fleshy tongue. I was able to explore new places and went more into the cheeks and down along the side of her tongue to the lower lip and so incorporated a larger territory of her mouth. I could feel her shift, her mood, and mental state change.[23]

It was Charles Darwin who first pointed out the similarity of facial expression across species. Like commonality in brain-body development and attachment, this observation suggests that the model of transference-countertransference might hold across species. Such analogies have been uncharted territory for modern science, but neuropsychology has brought the body back into the realm of the mind.

Once occupying four conceptual quarters, mind, body, animal, and human are reintegrating into one. Touch sparks a conversation between paired unconscious minds through the medium of the body. Bad dreams and horrific experiences that trauma seeds do not disappear by talking alone; processing conversation must be done in dialogue with the body.[24] The significance and meaning of Riesterer's work begin to appear, even through the lens of science, as more than simply "feeling good." Or, conversely, feeling good deserves a lot more respect than it usually gets.

In 2004 Elke was invited to the Thailand Elephant Hospital to demonstrate therapeutic touch and to inspire local veterinarians to use these approaches and this philosophy when working with elephants. Krungsee was a forty-year-old female, a captive logging elephant and victim of a landmine she had stumbled across during the course of her work. She was being treated at the clinic for her injuries. Used to equating human contact with painful disciplinary actions, the elephant first shrank away as Elke approached.

> Krungsee was in severe pain from her injury and the years of hard labor. Her routine at the hospital was a daily hobble down from her space on the terraced hill where she was chained each evening to the cement floor of the veterinarian station. She had become very disconnected from the rest of her body, most the time staring into space with a glazed look. As a result of holding up her leg for three years to compensate her injury, her muscles had atrophied. Her profound hurt showed when I moved my hand toward her stomach because she awoke immediately and reacted by lifting her rear left leg and pulling it forward protectively out of my reach, until, after a while, I approached and began to brush her with a soft broom. Little by little, her body softened, she sighed big deep breaths, eyes closed half shut, and tears ran down. When we finished the session, she coiled her trunk toward the back to touch my face that was also wet with tears.[25]

Touching bodies, touching souls.

11

Where Does the Soul Go?

According to the African Elephant Studbook, Medundamelli was born in Zimbabwe in 1969 and, given the year of her birth, almost certainly was captured during the mass killing of her family. At age two, probably while tethered to a dead body and after watching the ivory hacked off the faces of her relatives, "Dunda" was captured, and she arrived at the San Diego Wild Animal Park on September 18, 1971. On February 4, 1977, she was transferred across town to the sister institution, the San Diego Zoo, then brought back to the Wild Animal Park, near Escondido, California, a year later. The head zoo elephant keeper, Alan Roocroft, led four men with wooden ax handles who beat Dunda for two days to "discipline" her while her legs were chained so that she was nearly prone. Only when she acquiesced to their commands and accepted an apple did her punishment cease. The San Francisco zookeeper

Paul Hunter was quoted at the time as saying of elephants, "You have to motivate them and the way you do that is by beating the hell out of them."[1] *The San Diego City attorney would not press charges because "evidence [was] at best inconclusive" and the disciplinary technique was "accepted."*[2] *The U.S. Department of Agriculture issued a reprimand and levied a fine, but Roocroft and the others were cleared. On December 9, 1993, Medundamelli was moved to the Oakland Zoo, where she now resides.*

It is difficult to think of any other species that evokes such contradictory behaviors and attitudes in humans throughout their history than the elephant.[3] We have seen elephants worshiped, slaughtered, tortured, nurtured, admired, feared, hacked, rescued, studied, caressed, symbolized, and eulogized. Something remains unresolved and unsolved: why? Why, even now, when scientific theories and data declare that elephants and humans are comparably vulnerable to suffering, do humans continue to objectify and inflict pain upon their pachyderm kin?

The contradictions that humans express about elephants may stem from a sense of the incongruities that elephants embody. Elephants are huge, powerful, and so strong that a slight flick of a trunk or leg can mean instant death to a person standing by. Yet they are benign in their vegetarian habits and peaceful restraint, and they rarely exert violence. But this still does not explain why people behave the way they do with elephants.

Consequently, we turn from the exploration of elephant minds to those of humans. The fate of elephants itself depends on the human mind, for what, how, and why we think the way we do has had a huge impact on elephant lives. Perhaps by understanding humanity's contradictory behavior toward elephants, we can find out something about ourselves and why, on one hand, we admire elephants, spending time and money to study, watch, and care for them, yet we are willing to subject them to conditions that threaten their extinction and cause undue pain.

Contradictions are perhaps nowhere better illustrated than in the story of one of Elke Riesterer's long-term clients, Medundamelli, formerly of

Zimbabwe. Also known as Madunda-Mela, Madunda, or Dunda, she was the subject of a notorious court hearing concerning her treatment in 1988 during the time of Ray Ryan's tenure at the San Diego Wild Animal Park. Dunda's maltreatment might never have been exposed had not some of the zoo's personnel brought the event to light and the media taken an interest.

While television is replete with gruesome details of humans injuring and killing other humans, the agonies of an animal somehow evoke an even greater repugnance. Perhaps it is intuitive recognition that animal victims are only bystanders made casualties of a purely human problem. Dunda's story is one of many similar incidents involving elephants, but because the details were publically documented, they provide a rare portrait of violence that can help clarify the plight of elephants and the motivations of the people and institutions involved.

As is true of many emotional experiences, details of the case are in dispute. Discrepancies surface even concerning dates when Dunda received her life-threatening beatings. A news release from the office of San Diego City Attorney John Witt states that the incidents occurred February 18 and 19, whereas the dates referred to in other reports, including a letter from A. J. Wilson of the Animal and Plant Health Inspection Service (APHIS), responsible for investigating and enforcing the Animal Welfare Act (AWA), are February 17 and 18. Similarly, suspects' names are variously spelled. Nonetheless, despite some factual inconsistencies, there is overall agreement about what happened. Dunda was subjected to shocking brutality at the hands of the San Diego Wild Animal Park personnel in charge of her care.

Dunda is described as very insecure and easily frightened. In 1982 she fell into an exhibit moat and fractured her jaw. Three weeks later, when keepers noticed a suppurating wound on the left side of her face and bone chips in her feed, Phil Ensley, a wildlife veterinarian with the San Diego Zoo since 1976, was called to examine her. X-rays revealed that a significant piece of her jaw had broken off, and more serious fractures were suspected.

After administering a comprehensive regimen of antibiotics and cleaning the wound, Steve Garfin, then an assistant professor and now chairman of the Department of Orthopaedic Surgery at the University of California, San Diego, Medical School, was contacted to contribute his expertise in human oral injuries. The doctors suggested attaching a metal plate to help hold the fractured jaw in place and to avoid cracking the lower molar, which is essential for elephants to chew their food, but Ensley's experience suggested that without the ability to administer antibiotics continuously, as would be done for humans, the implant could lead to more serious infections. Dunda's prognosis looked iffy. But despite several anesthesia procedures and incidents of infection, necrosis, and immunological failure, she managed to pull through. Several months later, Ensley shared the joy of other successful physicians: witnessing recovery and appreciation. When he visited Dunda in the exhibit, she reached out her trunk and put it in his hand. Six short years later, however, her fortune changed again.

In February 1988 Dunda was returned to the Wild Animal Park. The curator of mammals, Lawrence Killmar, said that there were two reasons for the move: first, the zoo was overcrowded, and second, Dunda was a difficult elephant. Investigation revealed that she had sustained two days of brutal beatings starting the day after her arrival at the park. Killmar told the investigating officer that "he did not believe there was any cause for alarm in this incident": Alan Roocroft, he said, was "an excellent trainer."[4] After the news became public, a formal investigation was made and a hearing scheduled.

Testimony revealed that while the treatment of Dunda was shockingly violent, it was far from unusual. A portion of the exchange between the hearing chairman, state senator Dan McCorquodale, and one of Dunda's assailants and Ryan's former boss, Lou Bisconti, illustrates the point:

CHAIRMAN MCCORQUODALE: Have you disciplined any of the other elephants since you've been there at the Wild Animal Park?
MR. BISCONTI: You mean in the manner. . .
MCCORQUODALE: They're all older, aren't they?
BISCONTI: In the manner of the way that Dunda was disciplined?
MCCORQUODALE: Yes, right, yes.
BISCONTI: Yes, I have.

McCorquodale: After those sessions, were there similar marks on the elephants as Dunda or was there. . . ?

Bisconti: Yes, there were.

McCorquodale: So the sloughing away of skin, that type of thing, was not unique. You told the San Diego Humane Society Investigator that if Dunda had been abused she would be schizoid. What do you mean by that?

Bisconti: Well, we would have driven the animal off the edge. We would never be able to approach her. See, we're trying to achieve a relationship with the animal where we're going to be able to get around her, we're going to be able to work with her. We have to have a close relationship. She has to at one point to start to like your presence. And when she starts liking your presence, then you can get around her, you can do things with her, she starts to trust you.

McCorquodale: Is that generally a good test, do you think, to determine if actual abuse has occurred? Do you think that—is that the general view?

Bisconti: I think you can drive an animal off the edge. Yeah.

McCorquodale: And not going to that point, is it a determination that abuse hasn't occurred? Or could abuse occur and still not reach that point? You really may not have that background to say.

Bisconti: I think the way Dunda is responding today, with her keepers today, is proof that we didn't drive her over the edge and that we didn't abuse her.[5]

George "Slim" Lewis, whom we met earlier during the story of Black Diamond, describes a disciplinary practice similar to that which Dunda underwent, except that he employed a steel automobile axle. Lewis admits that he lost his temper in seeking to subdue the bull elephant Ziggy, "one of the biggest wild-eyed monsters" he had ever seen. "Under such circumstances, a man in a temper is like an angry boxer. He's likely to forget his caution and get clipped. But also, like the boxer, he may land a quick knockout punch before that happens. I was lucky. I sailed into Ziggy in a fury, and fifteen minutes later he was lying on his side on the floor." Lewis also speaks about the use of chaining. "Sometimes it is possible, when the elephant is chained front and rear to win without whipping it, provided the man has more patience than his elephant. It's done by gradually tightening the chains, a few inches at a time, until the elephant is supporting its weight entirely on the front and hind legs that are free."[6] Circus and many if not most zoo trainers believe that AZA-approved physical punishment is necessary to control elephants because, they maintain, the same behavior is found in free-ranging elephant culture: elephants, they say, discipline subordinate members, an assertion that is disputed by elephant ethologists.[7]

Routine or not, Dunda's beating was startlingly brutal. Steve Friedlund, the senior animal trainer at the zoo, reported what Bisconti had told him:

> I asked how badly Dunda was hurt, and [Bisconti] replied, "Well, she rolled over on her side and moaned." I was appalled when I heard this and I still can't get it out of my mind. Lou described the area around Dunda as covered with the horrific flow of nearly liquid feces. This condition of excessive elimination is a sure sign that the animal was in a state of fear and is easily recognized. An animal being disciplined for aggression would not do this until it had been frightened by the disciplinary action and would understand what was happening. Dunda was defecating and urinating profusely from the very beginning. . . . This frightened, cornered animal was in a state of panic situation and had no idea what was happening to her. She was fighting for her life, and she must have felt it was over.[8]

Lewis also mentioned a similar reaction by elephants when disciplined in the circus.

Dunda's wounds provided further evidence. According to Friedlund, photographs of one wound showed a "deep crevasse in the left forehead . . . so deep that it reminds you of the time Alan Roocroft described the use of sledge hammers to discipline elephants in Hamburg." James

Baker, the state humane officer assigned to the investigation, reported: "Dr. Ensley said that the wounds were caused by blunt [force] trauma. He does not believe the wounds resulted from cage trauma or Dunda's attacks against the bars." In his opinion, Ensley told Baker, Dunda's discipline "went too far, the wounds were too extensive"; he said that he "had never seen wounds that bad on an elephant and he never wants to again." Baker said Ensley went on to comment "the whole episode bothered him. It made him sick." Cleveland Amory, the cofounder of Fund for Animals and president of the New England Anti-Vivisection Society (NEAVS), stated: "I have been appalled since I've listened to these hearings this morning that a very fine zoo in a very fine state, probably the most caring animal state in the United States, with a very fine Legislature, should allow itself to get in this position."[9]

According to Ray Ryan, Roocroft was intolerant of anyone who did not do exactly what he wanted done when he wanted it. Dunda was nervous and defensive after her abrupt move from the zoo to the park, and when she failed to respond to Roocroft as he wanted, the beatings commenced. The court records show that most of the five men were between their late twenties and early forties, and that four men were between five-foot-eleven and six feet and weighed 170–200 pounds. In other words, all were strong, fit, and capable of dealing powerful blows against a chained elephant. There was also a sixth man, Bruce Upchurch, a visiting elephant keeper from the Woodland Park Zoo, Washington.[10] He denied hitting Dunda and stated that he had only "worked the chains and ropes." Upchurch "did not see anything that shocked or upset him" and said that "when necessary, the Seattle Zoo uses the same type of discipline."[11] Ryan describes the incident differently:

> Dunda was all freaked out. Ask Patty Hearst how she felt. Taken away from her group, out of the blue. Roocroft was pissed that she wouldn't do what he wanted and that she was acting hostile to him, so he and the other guys hooked up her back legs with chains and the front with a three-foot give so they could pull her down to the ground with a block and tackle

> using comealongs from a truck and beat her almost continually over the next two days. Just because she wouldn't listen to Roocroft they had to bring her down. Four, five guys on either side beating the living crap out of her till she fell over even in chains. Two days of beating five hours each day. The skin on the left side of her head fell off and her urine had blood in it.[12]

Despite overwhelming evidence of cruelty and damage that Dunda sustained—and the distaste exhibited by public officials and public at large—the San Diego attorney's office announced that no criminal complaint would be filed, and no punishments were levied against the five men named as assailants, Alan Roocroft, Louis Bisconti, Patrick Humphrey, John Crayon, and Edgardo Marqués, or against the Wild Animal Park. Roocroft is no longer with the park, but is still working in the industry.

Ryan is honest about his own violent behavior on the job in San Diego. When he first arrived at the Wild Animal Park, he was shocked at the dramatic contrast with his previous work with primates and the physical abuse meted out to the elephants. He had changed radically since his unruly teenage days in Chicago:

> I was very confrontational there because of the way I grew up. I really liked to fight. I used to be tough—like the guys in *West Side Story*. We'd go out at night and jump out and smash car windows just 'cause we were bored. But working with the monkeys changed that. I gave up all that tough stuff and went to the beach and hung out and talked with the primates and it made me realize how mean I had been back home. And then when I came to the Animal Park I got mean again, and mean to the elephants. It happened sudden-like when one day Peach kicked me with her back left leg and threw me 50 feet in the air, blew out my shoulder. The other elephant guys told me to beat her. And I did, not as bad as the others, but it was scary

> because I found all this anger again. I didn't know there was all this I could still draw from.[13]

Today he reads the works of the Dalai Lama, speaks out about violence against elephants, and advocates against elephant captivity. Ryan's temperament is calm, as it was during the days when he communed with monkeys. He is wistful about his time with the elephants, paraphrasing a song to voice his regrets: "I wish I could hear all those stories when I couldn't listen." He acknowledges that some zoos have changed to a degree, but not all, and not enough, and circuses not at all.

While Dunda's case raised the visibility of captive industry abuse, the same practices continue. A coalition of animal protection groups have sued Feld Entertainment, owner of Ringling Brothers and Barnum and Bailey Circus, for systemic abuse of the company's fifty-four Asian elephants who are protected by the Endangered Species Act. During the federal trial, videos showed the handling that circus elephants receive and the injuries they suffer. Owner Kenneth Feld admitted to the routine use of force and instruments such as the ankus.[14]

Despite the notoriety of Dunda's case, another incident of egregious mistreatment surfaced twelve years later, again in a prestigious zoo, this time in Portland, Oregon. A senior Oregon Zoo elephant keeper, Fred Marion, allegedly came to work under the influence of alcohol and, without provocation, chased Rose-Tu, a five- or six-year-old female elephant, around the yard shouting, swearing, and hitting her.[15] Two assistant keepers confirmed that Marion then stuck an ankus into Rose-Tu's anus and pulled hard down on it twice, after which she crumpled to the ground in pain. Finally, other keepers and managers were called in and stopped Marion's rampage. Evidence of at least 176 wounds were counted on Rose-Tu's body. She was scarred in other ways as well. Later mated with a bull elephant, Tusko, on loan from another zoo, she gave birth to a 286-pound male baby. Within hours of his birth, Rose-Tu trampled him. The infant elephant was removed and examined and showed no apparent injuries. Rose-Tu was chained, gradually reintroduced to her baby, and eventually began to nurse him.

It would be untrue and inaccurate to see everyone in the captive industry as sadists or even insensitive. Nor is the treatment of every elephant in captivity as extreme as Dunda's. As we saw in her case, some individuals draw the line when ethical standards are breached and are profoundly disturbed by abuse. Certain zoos have worked intensively to better the conditions for their animals. While considering captivity an unethical institution, some staff members stay on to minister to the elephants as best they can. Many keepers and others who help care for elephants express a deep love for those they watch over. Indeed, it not uncommon to hear diverse people in both the circus and zoo industries speaking emotionally about elephants.

"Slim" Lewis, whom we heard describe the physical excesses he imposed on the male elephant Ziggy, years later commented, "Yes, I cried when I heard the news" that Ziggy had died at age fifty-eight. "My weeping over the death of the 'meanest elephant in captivity,' one that tried to kill me, made 'odd' news items across the country. But I loved that rogue."[16] Kenneth Feld, whose enterprise is openly criticized for its brutality, has been described as someone who "talks like an elephant lover."[17] Zookeepers and caregivers speak about how much they love the elephants and at times stay up all night caring for an ailing animal. When an elephant dies, tears are shed. Caregivers watch in grief while elephant comrades mourn a former exhibit-mate. When a young Woodland Zoo elephant died, zoo president and CEO Deborah Jensen stated in a press release: "Our beloved Hansa's short life was in the best hands of elephant care and management. Of course, our elephant care staff, other staff and volunteers, and our Board are deeply saddened. We will all miss her. We are a family here and we will provide as much support and comfort as possible during this difficult time."[18]

Similarly, a Houston Zoo spokesman eulogized two-year-old Mac: "The entire Zoo staff is absolutely devastated. It's difficult to put into words and describe the attachment that has developed over the past two years. This is a terrible loss especially for the elephant keepers and veterinary medical staff who cared for Mac during his time with us." At Zoo Atlanta, where Dottie, a twenty-six-year-old pregnant with her first baby, died, president and CEO Dennis Kelly commented: "We are saddened by the sudden loss of Dottie." Invariably, staff members ascribe human qualities to elephants and declare that this kinship brings the two species close together. "They'll grieve just like people will. They're incredibly intelligent animals," said Houston Zoo director Rick Barongi, commenting on the other zoo elephants' behavior following Dottie's death.[19] There are even zoos that have erected memorial sites.

Evoking something of what Tina Turner sings, "What's love got to do with it?" Carol Buckley acknowledges the love expressed by workers in the captive industry, but says: "I've known people in this business for thirty years. I know they love elephants. What I have had to learn to understand is you can love someone in a very dysfunctional way."[20] Buckley is among a growing list of individuals (Ray Ryan, the late Les Schobert, and Pat Derby are some others) who changed their lives for love of a different kind. They left the industry, aware that they would suffer the consequences financially, professionally, and socially, to become vocal elephant advocates. Some veterinarians, such as Mel Richardson, retain their medical practices but speak openly about the harm they see done, and decry the brutality of captivity.

It is obviously somewhat puzzling that so many people can spend weeks, years, even lifetimes with elephants, acknowledge their extraordinary sensitivity, intelligence, empathetic and ethical qualities, and nurturing ways, even develop deep bonds with them, yet continue to perpetuate elephant suffering. As Ray Ryan put it poignantly, "Why don't we just let them be who they are?"[21]

The practice of holding animals captive has a long history.[22] Even while acknowledging historical complexities, our primary concern here is understanding the mechanisms that preserve present-day contradictions that allow elephants to be kept captive despite the scientific knowledge about them. Our goal is to identify what factors today, not in the distant past, make it possible to recognize and deny human-elephant comparability. How does one weep at the passing of a dear zoo resident who has been regarded as intelligent, sentient, and equipped with humanlike emotions and values, yet knowingly watch her wither day by day?

This question is not rhetorical. It relates to a very real and important phenomenon that affects humans: psychological dissonance, a mental state that permits knowledge of one reality and its simultaneous denial. Denial is a subject that has preoccupied many writers. Susan Griffin is one.

In *A Chorus of Stones: The Private Life of War,* Griffin describes how denial is woven into the privately experienced reality. "There are many ways we have of standing outside ourselves in ignorance. Those who have learned as children to become strangers to themselves do not find this a difficult task. Habit has made it natural not to feel. To ignore the consequences of what one does in the world becomes ordinary. And this tendency is encouraged by a social structure that makes fragments of real events. . . . For some, blindness becomes a kind of refuge, a way of life that is chosen, even with stubborn volition, and does not yield easily, even to visible evidence."[23] Closeted family skeletons, unacknowledged grief, and little white lies make up Griffin's chorus of stones. These secrets may be temporarily silenced, but in time they distort the lives of future generations.

The psychiatrist Robert Jay Lifton has also dealt with the topic of denial. His work focuses on the history of the causes and effects of violence associated with atrocities of the Second World War. While the majority of traumatological research focuses on victims of violence, Lifton's subjects include perpetrators. Through intensive interviews with Nazi doctors who had staffed Nazi Germany's concentration camp Auschwitz-Birkenau, near Krakow, Poland, Lifton's psychohistory explores how professional healers, doctors trained to save lives, participated in an agenda of killing and sent so many to their deaths. He calls the psychological principle by which he describes the mental state and behavior of Nazi doctors "doubling."[24]

Denial, dissonance, and doubling are all psychological methods of survival. People use many tactics to adapt to conflict or other extreme conditions that create unhealthy stress. When physical adaptation (like running away from a charging bear) is not possible, then the mind is perhaps at its creative best as it resorts to diverse coping strategies. Each method of self-protection has a slightly different mechanism. Splitting occurs when the mind suspends the idea of a moral continuum, allowing a person instead to see himself as completely good and another as completely bad. This oversimplification of reality provides a rationale for extreme action, even killing. On the other hand, psychic numbing, according to Lifton, is "a form of dissociation characterized by the diminished capacity or inclination to feel, and usually includes separation of thought from feeling." Part of the mind becomes anesthetized in order to deal with what it cannot escape. Dissociation is "the separation of a portion of the mind from the whole."[25] In general, dissociation creates a kind of distance that allows someone to function, to do and say things that otherwise he could or would not. In war, for example, a soldier must perform actions he might formerly have regarded as unimaginable. But there is a cost to mental remodeling. Although dissociation provides escape from otherwise inescapable circumstances, symptoms of trauma and PTSD often follow when the terrorizing event is over and quiet descends.

Lifton maintains that, unlike multiple personality disorder (now known as dissociative identity disorder [DID]), in which the dominant self can vaguely recall his or her other selves, doubling does not involve complete mental separation, nor is it necessarily associated with childhood trauma. Doubling falls within the same category as dissociation, splitting, and numbing as a coping mechanism, but differs primarily because it involves the creation of two functioning wholes, not parts, within the self: doubling "carries the dissociative process still further with the formation of a functional second self, related to but more or less autonomous from the prior self."[26] Each self represents an opposite aware of what the other is doing and thinking.[27] Doubling does not preclude the simultaneous occurrence of splitting and numbing, but in Lifton's mind those are very different psychologies.

Five elements characterize this psychology. First, an interdependence exists between the two selves. They are both autonomous and connected. For example, the camp or "Auschwitz self" (what Lifton calls the doctor's self when working in the camps) needed to be autonomous, free of moral standards that applied outside the prison, in order to function in the demanding environment. At the same time, his other, "humane," self, the principled doctor and family man, could retain his integrity and function in the camp environment unburdened by its horrors. Second, the two selves are "holistic," able to integrate and remain coherent with their respective environments—the camp self was able to be a doctor performing selections for the gas chambers and the humane self was able to be the congenial host and father outside the prison. Third, while Lifton considers doubling to have evolved unconsciously, without the use of verbal language, a connecting awareness of the coexistence of the two worlds and selves results in a major shift in moral consciousness. This is illustrated in the case of Dr. Ernst B.

B. was keenly aware of the ethical schism between his Auschwitz self and his humane self. He attempted to avoid certain tasks, such as selections, and told Lifton that he had had "extreme 'inner resistance' to [his wife] visiting him" at Auschwitz. He acknowledged—without directly referring to himself—that "people were likely to reconstruct past events in a way that was favorable to them." Nonetheless, it was clear throughout the interviews that although Dr. B. expressed ambivalence, he also revealed "the extent of his Nazi involvement."[28]

A fourth characteristic of doubling is illustrated by the Auschwitz self's identification with survival. To live in the death-dominated camp environment—and thereby ensure survival of the humane self—the doctor had to create the "killing self." The final, and related, characteristic of doubling was guilt avoidance. The Auschwitz self was assigned to and solely responsible for doing the "dirty work." This allowed the humane self to retain his integrity outside the camp.

C. G. Jung evokes something like doubling (or splitting) in his discussion of the association between shared guilt and mass violence that occurred during the World Wars. He maintains that collective guilt, which he defines as "a state of magical uncleanness on the primitive and archaic levels," exists among all people since "no man lives within his own psychic sphere like a snail in its shell but is connected with his fellowman by his unconscious humanity."[29] Jung speaks of this relationship when describing the palpable sense of relief expressed by non-Germans at news of the 1933 burning of the Reichstag: "The Reichstag gave the signal—now we knew for certain where all the unrighteousness was to be found." This evil was apart from the rest of Europe because "we ourselves were securely entrenched in the opposite camp, among respectable people whose moral indignation could be trusted to rise higher and higher with every fresh sign of guilt on the other side. Even the call for mass executions no longer offended the ears of the righteous, and the saturation bombing of German cities was looked upon as the judgment of God."[30] Unconscious guilty recognition of what was happening in Germany provoked conscious distancing and splitting behavior.

Not all doctors exhibited exactly the same behavior. There was considerable variability and at times confusing contradictions among the men whom Lifton studied. For some, such as Dr. B., there was a certain ambivalence—the contradictions stark but disarming. For others, like Josef Mengele, the doubling was seamlessly clean. Lifton describes an event in which Mengele jumped from his car to save a Gypsy from drowning, then put the man back into the truck, which was on its way to the gas chamber.

Doctor Eduard Wirths provides yet another example of the variability of doubling. Wirths was the chief doctor of the Schutzstaffel (SS) and was considered by German staff and prisoners alike as one of the few irreproachable members of the staff, a singular man who was above corruption and characterized by the highest integrity. Wirths showed an obvious conflict within the self that could not be resolved. He was actively protective of prisoner doctors, but he ruthlessly objectified women and experimented on them. Lifton considers Wirths's final act dramatic evidence of internal conflict between the selves. After Wirths surrendered, a British officer approached and shook his hand, saying that he now knew what it was like to shake the hand of a man who had sent four million to their deaths. Within days, Wirths was dead, a suicide.[31]

Lifton's work is remarkable, if only for its unrelenting sense of presence. There is not one phrase in which he relaxes his attention and falls into academic remove. Despite clinical analysis, theory does not overshadow or caricature the named and unnamed individuals who were part of that particular cultural necrosis. As time passed and the Shoah became a universal reference, the Nazi doctors shed their mortal banality and grew into bigger-than-life archetypes of quintessential evil and wrongdoing, inexcusable trespassers on inviolable ethical ground. However, historical remove and mythical quality tend to dissociate the horror of the past from the present. Sitting only as close as the pages of a book, or even beside someone who has survived to tell, tends to make events remote, beyond present ownership, and part of someone else's life, not our own: "Distance from execution helps render responsibility hazy."[32] One is tempted to ignore that such extraordinary wrongdoing could happen or that it is still happening.

Can Lifton's psychoanalytic approach be used in other instances? Is doubling a viable lens through which the captive-elephant industry can be examined to explain, at least in part, the human contradictions surrounding these unique herbivores? Or are we guilty of seeing evil (and, one might add, trauma) everywhere, and do we thereby risk, as Tony Judt, warns, "losing the capacity to distinguish between the normal sins and follies of mankind—stupidity, prejudice, opportunism, demagogy, and fanaticism—and genuine evil."[33] Is it fair to compare the psychologies of those who incarcerate and objectify elephants and those who incarcerate and objectify people?

Such comparisons can evoke strong emotions and tensions. Memory of Hitler's regime stirs the archetypal: incidents of the period have become almost synonymous in the West with evil and unconscionable crimes. Indeed, it is difficult not to believe so, yet there have been other genocides warranting equal condemnation. Some, like the decimation of American Indians, lie restive but quiet. The American Indian genocides were less focused than that executed by the Nazis, but the deeds that brought North American indigenes and their cultures to near extinction are not so different, nor are the plaints of witnesses. Remniniscent of elephant elders' testimonies, an Indian man speaks of the devastation during his lifetime:

> My sun is set. My day is done. Darkness is stealing over me.
> Before I lie down to rise no more I will speak to my people.

> Hear me, this is not the time to lie. The Great Spirit made us and gave us this land we live in. . . . Then the white man came to our hunting grounds, a stranger. . . . They took away the buffalo and shot down our best warriors. They took away lands and surrounded us by fences. Their soldiers camped outside with cannon to shoot us down. They wiped the trails of our people from the face of the prairies. They forced our children to forsake the ways of their fathers. When I turn to the east I see no dawn. When I turn to the west the approaching night hides all.

An Omaha tribesman expresses the emptiness in the wake of violent attrition: "Now the face of all the land is changed and sad. The living creatures are gone. I see the land desolate and I suffer an unspeakable sadness. Sometimes I wake in the night and I feel as though I should suffocate from the pressure of this awful feeling of loneliness."[34]

A number of authors have discussed the historic parallels between the treatment of animals and that of humans in concentration camps, notably Charles Patterson in his book *Eternal Treblinka,* the title of which is taken from Isaac Bashevis Singer's story "The Letter Writer." Singer writes, "In their behavior towards creatures, all men are Nazis." Tellingly, too, Ray Ryan once commented, "When I worked at the Wild Animal Park, I used to say I was a guard at Auschwitz, and that's what we are."[35] But given the charged suspicion that can attach to comparisons made between camp guards and zookeepers, people and animals, human prisoners and elephants, it is important to lay out and review precisely what is being compared.

Science has established the comparability between humans and other vertebrates, including elephants, in terms of their capacities to be affected by confinement, isolation, pain, and psychological, emotional, physical, and social deprivations. In form and effect, then, human prisoners and elephants in captivity are equivalent, as are prisons and zoos. Prisoners in concentration camps such as Auschwitz are held against their will, as are elephants in captivity. Inmates in camps were exploited for their captors' benefit (in medical research experiments, for example), as elephants are kept in zoos and circuses for entertainment, profit, or even education: all have suffered from objectives tailored to values and needs of the dominant group. Within a limited circle, concentration camps became culturally acceptable institutions; the same is true in some cases of prisons for individuals who disagree with the state, and the same is true today of zoos and circuses.

One major difference between institutions is that Auschwitz-Birkenau was designed as a "death camp," a place where prisoners were intentionally killed. While captivity in zoos and circuses leads elephants to premature deaths, and some psychologists speculate that the dismal statistics of mortality and poor health among such elephants may be the consequence of unconscious psychological hostility, no overt program of killing exists.[36] Indeed, it is in their captors' interest to keep them alive so that they can continue to attract paying customers and visitors.

Another apparent difference is the assumption that—unlike Jews, Gypsies, and other victims of state persecution in Nazi Germany—elephants are simply not human, and thus are not accorded human rights. But it was precisely the Nazis' refusal to accept those "undesirables" as human—and the passage of infamous laws to cement that denial—that made possible, and technically legal, the extermination of six million–plus human beings. The Nazis' grotesque distortion of law and convention denied legal protection from abuse and murder to a vulnerable population. Here we have the opportunity to set a precedent of extending comparable protection to another vulnerable population that suffers from institutional behaviors and rationalization parallel to the Nazis. Given what we know about the science of elephant psychology, the extension of such protection would be not a distortion of law and convention but a fulfillment of their spirit.[37] Yet we do not acknowledge their rights as reified by science.

Lifton himself provides reasons to expand from his specific example to other cases. The brutality of the camps was not, he concludes, an anomaly resulting from a unique convergence of conditions to create a behavioral "perfect storm"; rather, the doctors' behavior seems to have presaged a psychology that would come to characterize modernity. Victimizer doubling can evolve in any type of social group, particularly those whom society invests with authority and whose "prior, humane self can be joined by a 'professional self' willing to ally itself with a destructive project."[38] Lifton deems physicists, artists, writers, psychologists, the clergy, the military, and biologists vulnerable to doubling. Today's prevalence of a professional identity separate from the private personal self prompts Lifton to ask: Is this the century of doubling?

"Indeed so," Jung might have answered. He maintained that the potential for legitimizing large-scale violence threatens whenever an individual transfers his rights and responsibilities to an invisible state or organization, as the doctors did. A person who merges his identity with a group necessarily adopts its principles and steps into the reality that the group defines, what Lifton called the Auschwitz environment. A person's sense of morality is then "measured in terms of dependence on the state."[39] The giving over to the state violates what Jung considers to be the core process for developing moral consciousness.

Individuation, the process of coming into selfhood, requires each person to decide what she believes and how she, not a set of social mores, thinks she should behave. According to Jung, "Ethical problems cannot be solved in the light of collective morality," and "without freedom there can be no morality."[40] Becoming a member of a group, a particular profession in this case, creates the first ethical tear because the individual's reality is supplanted with that of the profession.

The greater the hierarchical organization in society, the greater degree to which group qualities are pressed on an individual, the greater the anonymity of responsibility, and the greater the ethical confusion: "I was just following orders." An enforced ordinariness is installed, and the "unthinkable" act or belief is normalized through a shared identity. Doubling, then, is a two-step process involving, first, ceding ethical responsibility to a group reality and, second, psychological separation from experience.

This sense of ordinariness was exploited by camp personnel and created an almost hypnotic complacency among its victims. Anne Müller, whose relatives were killed in one of the camps, speaks of the everydayness that seemed to settle over the general population: "For most of the society, life was lived as if none of this was happening. People had regular jobs, concentration camp workers went off to work in the morning and came home at night to loving families, a home-cooked meal, a warm bed."[41] Indeed, these were ordinary people. It was the face of the man next door, the neighborhood policeman, or the railroad engineer that greeted the prisoners at the camp.

The MIT professor Lisa Peattie describes how the unthinkable became normalized even for long-term prisoners and camp staff: "Prison plumbers laid the water pipe in the crematorium and prison electricians wired the fences. The camp managers maintained standards and orderly process."[42] The same air of normality is found in weapons labs and nuclear arms research (including MIT's main military research lab), "where the 'unthinkable' is organized and prepared for in a division of labor participated in by people at many levels" and "complicity is obscured by the routineness of the work, interdependence, and distance from the results."[43]

From this point of view, doubling emerges as a clinical explication of Hannah Arendt's "banality of evil," which she first articulated while covering Adolf Eichmann's trial in Jerusalem. Arendt coined the phrase to explain how "people who carry out unspeakable crimes, like Eichmann, a top administrator in the machinery of the Nazi death camps, may not be crazy fanatics at all, but rather ordinary individuals who simply accept the premises of their state and participate in any ongoing enterprise with the energy of good bureaucrats."[44]

Notably, Lifton does not consider doubling a psychological disorder. Doubling did not disrupt; it made the wildly contrasting lives of camp doctor and family man coherent and liveable. Neither does he agree that Arendt's banality held for the Nazi doctors. He makes this qualification: "What I have noted about the ordinariness of Nazi doctors as men would seem to be further evidence of [Arendt's] thesis. But not quite. Nazi doctors were banal, but what they did, was not. Repeatedly in the study, I describe banal men performing demonic acts. In so doing—or *in order* to do so—the men themselves changed."[45] The doctors were no longer the banal men that they started out to be. The cloak of their profession was just that: a cloak under which, in Lifton's mind, evil dwelled. The phenomenon of doubling, therefore, does not exempt moral culpability.

By implication, both to answer Lifton's question and to achieve insight into those aspects of the human psyche that cause elephant suffering, we have to look beyond the extraordinary, beyond the complacency of euphemistic language and routineness, beyond words, professional authorities, and titles, and see the lived reality of Hannah Arendt's banalities as they are experienced by and etched into the bodies and minds of Dunda, Jennie, Girija Prasad, Ely, Massa, the young bulls, and countless others.

As with rationalizations for the death camps, or, it can be argued, any institutional project, the architects of captivity developed a rationale to justify the existence of the institution. In the case of zoos, the current justification was crafted more than one hundred years after zoos and circuses had come into existence. Randy Malamud, a professor at Georgia State University who has extensively researched the history of zoos, says that there has been a "greenwashing" of zoos that began "coincident with the popularity of environmental movements in the world at large—late 1960s, early 1970s, Earth Day, the 'crying Indian' public service ad, and so forth. Zoos have always tapped into popular sentiments. In the nineteenth century, it was imperialism, today it's conservation and education. Now, it is argued that elephants need captivity because they are endangered in the wild and zoos are the only place of safety, preservation, and assured procreation."[46] But there is more to the rebranding than meets the eye.

Environmental awareness and the precipitous drop in elephant numbers in Africa in the 1980s led to a curtailment of endangered species' importation, and captive industries were faced with the threat of losing ready access to the stream of baby pachyderms. During his testimony at the Roocroft hearing, Steve Friedlund spoke out against the San Diego Zoo's abusive commercial agenda: "Disciplining a working or performing elephant has no value to the cause of conservation or education. Shows and rides use elephants only as beasts of burden, clowns, or buffoons, And it's a big mistake for the public to think that elephants might enjoy doing it. . . . If you want to see an example of abject mental poverty in elephants, take a look at the disgraceful animal ride set-up at the San Diego Zoo. The fact that it is very lucrative is why it exists at all."[47]

The executive director of the Columbus Zoo and Aquarium in Ohio, Jeff Swanagan, spoke with candor about the new reality: "Everybody thinks the zoo is all about animals but it's not. This is primarily about the people."[48] Public interest in elephant exhibits and resultant ticket sales are sufficient for the Oregon Zoo to win, even in the uncertain economic climate of 2008, a $125 million bond election to improve its elephant facility. Funded by a 2006 bond, the Dallas Zoo will spend $25 million to $30 million on an expansion to 10 acres, of which Jenny will share 3.75.[49] Baby elephants are the biggest attractions of all. Reporter Mike di Paola describes the fervor with which a young elephant is greeted at the Rosamond Gifford Zoo in Syracuse, New York: "The crowd cheers as elephants Romani and her daughter, Kirina, execute knee bends, trunk curls and modest headstands."[50]

In addition to importation restrictions, the industry is dealing with another hurdle. An alternative to wild baby elephant capture was captive breeding. But "deaths [in captivity] are greater than reproduction successes, [and] there is concern for the viability of North American herds." The success rate of captive elephant breeding programs worldwide is poor because of infertility in bulls, undiagnosed female reproductive disorders, and herpes virus. African bulls account for only 10.2 percent and Asian bulls 13.4 percent of the Species Survival Plan (SSP) population.[51]

Zoo leaders consider their "elephant populations in peril" because of high mortality rates, low reproduction success, and importation limits. They seek to offset captive-elephant "extinctions," and their own institutional demise, through "a rigorous, centrally controlled breeding program that could ensure the establishment of a reliable long-term supply of endangered animals for zoos."[52] This vision, coupled with increased commercialization in habitat countries, that has prompted fearful protests that the elephant is on its way to being domesticated like cattle.[53]

The AZA's sustainable breeding programs was not always widely known. In 1988, at the hearing on Dunda's treatment, Cleveland Amory expressed concern about the San Diego Wild Animal Park and Zoo "going into the breeding of animals." He said that "nine-tenths of the zoos in the country sell their excess animals to hunting preserves. . . . Now, I'm not saying Mr. Alan Roocroft and his bunch are going to sell these baby elephants. But . . . to think that they are the answer to the African elephant poaching problem or the Asian problem is simply ridiculous. What are you breeding more elephants for? What are you going to do, sell them to other zoos?"[54]

Perhaps Amory asked the question rhetorically, but that indeed is the case. Osh, whom we met earlier, being prepped for breeding at the Oakland Zoo, is also expected to serve as a sperm donor for other zoos; captive breeding has expanded and reproduction research intensified throughout the country. The Ringling Bros. & Barnum and Bailey Center for Elephant Conservation in central Florida is "dedicated to the breeding, scientific study, and retirement of the Asian elephant." The Smithsonian Institution's National Zoo in Washington, D.C., has a thriving elephant reproduction research program that was enriched by a $135,000 gift from Ringling in 2006 to study reproduction (this on top of a $180,000 gift Ringling gave the zoo in 2005 to study endotheliotropic herpes viruses [EEHV], which have killed a number of elephant calves in captivity).[55]

Unfortunately for the industry, rationales for keeping elephants and other animals in captivity do not stand up to scrutiny. Captivity in zoos and circuses is unsafe and fails to preserve essential ingredients for a species: physical, social, psychological, cultural, emotional, or genetic characteristics.[56] The notion that captivity allows species to thrive is also belied by low fertility, infanticide, and transgenerational stress effects.

It is also contended that zoos provide a vital opportunity to enlighten visitors and educate them about conservation, that once they learn about elephants, zoo visitors are more likely to support the elephants under siege in their native habitats. Authors of a 2001 report undertaken by the AZA disagree: no evidence of measurable educational benefits on the public has yet to be produced by the zoos; furthermore, "zoo visitors' knowledge of conservation was characterized as 'superficial.'"[57] Six years later, a National Science Foundation–supported AZA report found that there were positive effects.[58]

But independent analysis proves otherwise. A separate study examined the methodology and inferential rigor of the AZA study. Researchers found "six major threats to methodological validity that undermine the authors' [of the AZA publication] conclusions" and "there remains no compelling evidence for the claim that zoos and aquariums promote attitude change, education, or conservation in visitors."[59] The same conclusion has been reached by others. A former AZA president conceded that the "top two reasons that people visit . . . zoo[s] is to spend quality time with their family and to have fun."[60] Despite the "cherished claim that its member zoos are important centers of conservation, education remains an unproven hypothesis."[61]

Since education and conservation do not bear up under examination as viable reasons for elephant captivity, it appears that economic benefits are decisive in keeping big, unwieldy creatures in small, unhealthful places. In light of the wretched existence that elephants in captivity endure, zoos emerge simply as instruments of institutional trauma. With or without the conscious recognition of its employees, zoos are businesses that thrive on animal suffering. Perhaps a difficult conclusion in light of what are obviously sincere efforts by many to do right by the residents for whom they care, but science and history prompt us to look behind the good intentions and beyond the pleasant surface to what lies beneath institutions of confinement.

Zoos and circuses—indeed animal-captivity institutions in general—have been rendered ordinary because they are socially sanctioned. Their ordinariness has cast an obscuring veil. The conditioned visitor's eyes do not register the silent violence embodied in concrete enclosures, barred and locked gates, and routine methods, called husbandry, used to subdue and control. Ray Ryan speaks from experience about the culture of violence in zoos: "It's hard to describe, but when you eventually get control over someone who has no natural control and is so big, well, it makes you feel big. It is a real display of machismo. . . . You could show you were a real man if you could beat down a big powerful animal. And I could always tell who had had a fight with their wife the morning or night before. We have not changed much since cave days. Men are still beating up women, still trying to run the world with domination. And if you notice, all the elephants we work with are females."[62]

Gender is a recurring theme for Ryan. His personal experiences and insights are consonant with those of ecofeminists who stress the strong linkages between sexism and violence against animals. Just as individual acts of violence against women are patterned after a world in which one in three is raped or beaten, typically by a family member or someone she knows, social ideas about species and how they are treated are similarly influenced: violation occurs through enculturation.[63] Virgil Butler, who worked many years in a chicken slaughterhouse, then quit to become a poultry advocate, credited the macho culture and the need to make a living as reasons for partaking in what he called "horrendous work." As part of the American male culture, Butler maintains, he was conditioned "from birth to turn off the emotions of empathy and compassion for 'lesser' creatures."[64] Statistics suggest that some of the men's violence may have stemmed from personal experiences: research has closely linked animal abuse with human childhood violence and trauma.[65]

It cannot be overlooked that the abuses of Dunda were committed by a group, and the brutal actions were consistent with institutionally condoned discipline. Eyewitness and Senate hearing testimony show that the keepers were fully aware of what they were doing. Lou Bisconti confessed to Friedlund that he and the others were "sick" about the beatings "the first night and had trouble sleeping." (Several months later, one of the accused told a zoo staff member that he still had recurring nightmares from the incident.) Nonetheless, ambivalent feelings and remorse did not stop the men from taking up their axehandles again. As Friedlund reported at the hearing, Bisconti "also told me that they had repeated the process all the next day." Steve Friedlund's comments on Dunda's absence of hostility when not feeling threatened suggest that their motive was not a lack of knowledge: "'The saddest thing of all,' Lou [Bisconti] told me, 'is that Dunda was calm and approachable only after being placed between two adult elephants, Peaches and Sabu. Why this wasn't done in the first place only point[s] out the two main weaknesses an animal handler cannot afford: ego and incompetence. When the desire to dominate clouds a person's professional vision, he or she is no better than someone with no ability whatsoever.'"[66]

Ray Ryan speaks about this awareness as well:

> Circus elephants—walking in single file down a street. Why don't they run away? It's simple. It's because they're dead. They are dead souls in circuses and zoos. The only way to get elephants who are so powerful to do what a human wants with just a flick of their hand, is to beat the soul out of them. I saw in it Peach, when I beat her, I saw her soul leave. And they come to zoos as just kids, less than five years old, less than chest height, then we starve or beat them or a combination of both to make them dependent on us. Not the way to treat animals—and no need to. Eles are tactile and nurturing. Sometimes, I would sleep right with the elephants. One time, I had fallen asleep and woke up with Peach standing straddled over me to make sure the younger eles wouldn't step on me. They don't kill unless they are pushed to a point. Do you know how hard you have to push to get an herbivore to kill?[67]

What nurtured the repeated abuse by the San Diego workers? Could they have acted in fear of Roocroft, their boss? Perhaps. People compete and do nasty things to get ahead, or to keep what they have: "Normalization of the unthinkable comes easily when money, status, power, and jobs are at stake. . . . Companies and workers can always be found to manufacture poison gases, napalm, or instruments of torture, and intellectuals will be dredged up to justify their production and use. The rationalizations are hoary with age: government knows best, ours is a strictly defensive effort, or, if it wasn't me somebody else would do it."[68] However, thankfully, not all succumb to the unthinkable.

Many in the captive industry who work with elephants reflect Arendt's banality: they are decent, conscientious, warm, caring citizens with friends or families; they are not physically violent, and they volunteer at charities and school events. It is difficult and disturbing even to associate such individuals with the idea of "evil," and to charge them with being aware of the suffering that they indirectly or directly inflict.[69] Even the available testimony and data suggest that the behavior of Dunda's assailants is more consistent with splitting or numbing than with doubling. The angst and recurring nightmares of Dunda's tormentors speak of trauma. All of which, while not excusing those who make a living by working in zoos or circuses, redirects attention from individuals to the psychological architecture of the captive industry itself, which normalizes abusive behavior, encouraging someone like Ray Ryan to function "like a guard at Auschwitz" and the San Diego city attorney to pronounce Dunda's treatment as acceptably routine.

An internal memo of the San Diego Zoological Society Administration, signed by a Doug Myers, provides insight. It is dated March 26, 1989, a full year after Dunda's assault and is addressed to "Zoological Society Employees." The memo advises that all contacts with media or others resulting from a forthcoming *Parade* magazine article be directed to the zoo's public relations department:

> This Sunday, the weekly magazine PARADE will carry an article titled "Are Our Zoos Humane?" To illustrate the contention that zoos may not be humane, the author makes several references to the discipline of Dunda, an elephant transferred from the Zoo to the Wild Animal Park last year. Dunda's disciplinary trainings has been thoroughly investigated by the Zoological Society, by the San Diego Humane Society, by the San Diego City Attorney, and by the American Association of Zoological Parks and Aquariums, and these studies have found no wrongdoing by our Wild Animal Park elephant keepers. . . . We work for a proud and accomplished organization. It is a shame that a vocal minority can so easily mislead the media and public as they have with PARADE. In the 13 months since Dunda's transfer to the Park, we have hatched the first captive-

> conceived California condor, sent Przewalski's horses and Aribian [*sic*] oryx to their countries of origin for release into the wild, provided support for conservation work the world over, and numerous other positive accomplishments. Most important, Dunda is doing well at the Wild Animal Park, where she receives the best care and attention. Even PARADE lists us among the best zoos in the nation. Thanks to your continued fine work, that is a status we will long maintain.[70]

There are several messages implicit in this communiqué. One is that good deeds make up for bad: support for conservation makes Dunda's treatment acceptable. This type of rationale is echoed in Lifton's description of a physicist in the nuclear arms industry. It is "his commitment to democracy and family life [his humane self] that enables him to claim similar humanity for his nuclear weapons self [Auschwitz self] despite its contribution to devices that could slaughter millions of people."[71] Translated to elephant workers: a zoo administrator or keeper's commitment to care for elephants (humane self) enables her to claim similar humanity for her zoo self (Auschwitz self) despite its contribution to elephant suffering and death. The self sidesteps responsibility and guilt, as Otto Rank points out, not by thwarting conscience but by transferring the actions to another self, one whose criteria for behavior are alien to those of the humane self.

This brings up a second implicit message of the memo. The zoo and park do good things, therefore they cannot do bad things. Problems come from misinterpretation by outsiders. The zoo conserves elephants, and the outside "vocal minority" (in reality, a huge segment of the public, government institutions, and elephant professionals far outnumber zoo staff) is raising false issues. Good and evil are split off, parsed and assigned to different bodies: zoos and an unhappy public, respectively. This mechanism resonates with Nazi propaganda: it was the Jews who were the problem, who were to blame, and who constituted a burden to the regime.

When something bad does happen, it might be called regrettable but relegated to a minor detail lost in the grand fabric of the master narrative. Like slavery and the Vietnam War, Dunda's experience was "a tragic error . . . rather than a rational institutional choice; it has been marginalized rather than recognized as a central and integral feature . . . and it has been portrayed as an error being rectified in an ever-evolving process, rather than as a terrible permanent scar."[72]

Third, the memo suggests that truth is created and defined from the internal perspective of the zoo self. The zoo self is the authority on reality because it is sanctified by its society, thereby retaining a justified insularity. The zoo claims that it was exonerated, yet an official letter from APHIS signed by A. J. Wilson states that: "While we recognize the need for control and discipline of animals both to assure safety of the animals and the public, the physical harm and stress applied in this case were excessive and went beyond what APHIS considers acceptable disciplinary measures. Therefore we must warn you that any further documented use of such excessive disciplinary measures may result in legal action."[73]

The San Diego city attorney becomes complicit in maintaining the zoo's reality when he states,

> People who care about animals are and should be shocked at the prospect of a world-famous animal sanctuary inflicting needless pain on helpless animals in its care. . . . But the evidence demonstrates that didn't happen. There is no question that Dunda was struck repeatedly by Wild Animal Park elephant keepers. There is also no question that the hitting with axe handles and the stick end of elephant hooks caused trauma to Dunda's head. But our review . . . shows that the incident is far less serious than was first thought, and arose from a legitimate need to discipline and train a dangerous, four-ton elephant. . . . Our evidence is, at best, inconclusive as to the severity of Dunda's injuries. . . . Whether [Dunda's discipline] was excessive when judged by a professional standard

> we leave to the professionals. What is clear is that it cannot be said to be excessive when judged by a criminal standard.[74]

Given the copious amount of testimony during the hearing by a spectrum of professionals, including veterinarians, who judged her treatment excessive and her injuries traumatic, the convoluted logic and statements that blatantly contradict documented evidence are mind-boggling. The message appears to be that no matter what happened, no matter the facts, it was okay.

The zoo maintains its version of reality in other ways. By asserting that brutal discipline is typical in wild elephant society, the industry's disciplinary actions blend into the fabric of folklore peppered by selective science. When a reporter asks the Pittsburgh Zoo elephant keeper Willie Theison how a lack of space to roam affects the zoo's elephants, Theison replies: "Well, we provide everything they need so they don't have to roam." And when prompted to comment on long-term effects of culls on elephant behavior: "I don't buy that at all. I've spent too many years with too many elephants to think that would be the case." Questioned as to why one of the elephants, Moja, killed her handler, he answers: "We will probably never ever know." Even though scientific explanations exist for her aberrant behavior and that of other elephants, substantive data and knowledge are dismissed. "Perfect replication of nature in zoos is . . . undesirable." Zoos "should not be distracted by simplistic zoo-wild comparisons."[75] At the same time, attention is diverted from the captive elephant's plight with claims of conservation and promises to a public eager to connect with the mystical elephant and those who enviably stand so close.

Theison denies being a proponent of violence against elephants, while at the same time acknowledging that his friend and mentor was Allen Campbell, who was killed by Tyke. At the time of his death, Campbell "showed cocaine and alcohol in his system . . . [and] four years before that, he was fired from a job at the Denver Zoo after fellow employees accused him of abusing elephants."[76] The power of myth and illusion cannot be overestimated.

Lifton quotes testimony of a camp survivor who was incredulous about the power of delusion she witnessed in Auschwitz: denial so potent, so psychologically effective, that it extended even to the victims. The death camp administration erected false house fronts to mask and distract from the crematoria: "The houses that the crematoria had—you know, brick houses, windows, curtains, white picket fences around the front. And people never thought of anything—regardless of the chimneys smoking, They could not believe it."[77]

Researchers of totalitarianism have noted that institutional success is enhanced by the elimination of witnesses either through extermination, blocking access to information, or creating a threatening veil of secrecy. Obtaining information from the captive industry is possible only if the zoo is public, and as noted earlier, access to these records is often difficult and can be legally compelled only rarely by court-ordered subpoena. Even in public institutions, as we saw in the San Diego Zoo and Wild Animal Park, the effort to create an alternative reality provides a surrogate means by which witnesses are "eliminated." Elephant "voices"—scars, wails, infanticide, stereotypy, depression, pain, and loneliness—are silenced through denial, misinformation, and dismissal. The fabrication of reality and the zoo self define the sufficient conditions to make possible and encourage an individual like Ray Ryan to employ violence. Domination of elephants is normative and requisite to professional membership, and animal suffering in captivity disappears in mythological quicksand.

Inside the zoo and circus, reality is shaped to fit institutional goals: domination and control. Dunda's and other animals' unthinkable experiences are normalized through the banality of routinization. Violence is euphemistically coded as elephant "management" and "husbandry." Lifton makes an important point in this regard. The reality of what the camp doctors did was not denied, but its *meaning* was. The task of effecting prisoner selection, for instance, was not denied, but its meaning as a method for deciding who lived and who died, was. The doctors' reality was their engagement in the selection process, not as actual executioners of lethal action. In parallel, the use of violence against elephants as an integral part of their care is not denied, but the meaning of violence—as a brutal, cruel method to debilitate an elephant so that she will do what a human wants—is. Elephant captivity is not denied, but its meaning as a process of suffering and slow death is. The process of captive breeding (forced acquisition of semen, insemination, forced removal of the newborn elephant) is not denied, but its meaning as a method of eugenics is. The zoo self is right because of what the institution or group represents, not what it is actually doing, and so forth.

Psychologically, institutional doubling permits simultaneous existence as a humane and ethical member of society and as a sanctioned professional performing duties that would be unethical in other circumstances. The elephants and the human keeper are manipulated to keep, as Lifton describes the camps, the atrocity-producing situation running. Death camps and institutionalized elephant incarceration are possible because "they succeeded in making use of this form of doubling for tapping the general human moral and psychological potential for mobilizing evil on a vast scale and channeling it into systematic killing [or subjugation]."[78] These kinds of jobs and tasks are not arbitrary. Human trauma, the Harvard cross-cultural psychiatrist Arthur Kleinman reminds us, is a planned, intentional outcome employed to systematically silence through suffering.

But we cannot stop at the gates of zoos and circuses. The economist and media analyst Edward S. Herman concludes that the unthinkable is perpetuated because "there is also the retreat to ignorance, real, cultivated, or feigned. Consumer ignorance of process is important."[79] The myths of the Pittsburgh Zoo, the San Diego Zoological Society, and the captive industry as a whole thrive only because their listeners believe, or wish to believe.

Institutionalized captivity remains a tenacious barnacle well after the cultural tide has retreated because the culture in which they are embedded is complicit and guilty of doubling itself. Francis Jeanson, the activist who spurred on French intellectuals to speak out against the Algerian colonial war, wrote scathingly about the powerful role the public bystander plays in facilitating oppression and misdeeds they claim to abhor: "And if, apparently, you manage not to soil your hands, it's because others are doing the dirty work in your place. *You have your henchmen,* and all things considered, you are the real guilty party; for without you, without your blind indifference, such men could not undertake acts that condemn you as much as they dishonor them."[80] Elephant suffering does not lie at the feet of one culprit. Jung wrote: "The murder has been suffered by everyone, and everyone has committed it. We have all made this collective psychic murder possible."[81]

As Lifton tragically predicted, doubling has become a way of life, permeating perception, thought, and science, and when such violence is normalized to the extent that it becomes an accepted part of a celebrated cultural symbol—embodied by circuses and zoos—"only extraordinary people can resist the gravitational pull of an atrocity-producing situation."[82]

And what of Dunda? Here is a recent excerpt from Elke Riesterer's diary after a visit with Dunda, now M'Dunda, at the Oakland Zoo:

> Delighted by the juicy orange and tongue massage, she squeezed her eyes almost shut. Her pleasure was obvious and with each repetition she slipped deeper into that dreamy state of mind leading to greater relaxation and comfort. She moved her head to present her eyes to me, then she lowered her head so that I could reach the top of her head. I worked on her enormous delicate ears and then she moved again so that I could TTouch both sides of her body as well as her buttocks. She relaxed further and leaned against the fence structure. I did a number of Python Lifts, straight acupressure mixed in with TTouch. She relaxed her right hip by letting it slump down. Twice during the session she handed me thick tree sticks, the size of my forearm. After ten years, we had reached yet again a new depth of communication between us.[83]

Ray Ryan plans to visit Dunda this year.

12
Beyond Numbers

Chalakka Mohanan was a fifty-three-year-old Asian elephant, old enough to be a grandfather or even a great-grandfather, at a time of life to be surrounded by his progeny and looked up to by the young males in his group. But as a young elephant, he had been taken into servitude. When townspeople found him, he was alone. He had been abandoned by mahouts and left tethered with no shelter or food. Chalakka Mohanan, descendant of a proud heritage, starved to death. He was owned by Anwar Saddath of Perumbavoor, India, and had been brought by "middlemen" to parade in temple festivals. It was reported that the mahouts had tortured the tusker while drunk, then left hurriedly when they realized he was dying. A picture shows him fallen onto his right side, emaciated. His body was burned and his tusks were taken into custody by the Forest Department.[1]

Albert Einstein once said that theory decides what we observe, and this is certainly borne out in the case of elephants. Theories emerging from neuroscience and psychology have introduced a brand new way of seeing elephants, ourselves, and all other animals. When we read the Sheldrick Elephant Keepers' Diaries and Carol Buckley's own, they are more than charming just-so stories. By probing beneath antiseptic statistics of rhinoceros assaults, mahout deaths, and developing brains, science reveals something of what orphaned infant elephants and swaying captives experience. The new field of affective neuroscience has reconnected heart and mind and in so doing provides a formal way to explore what elephants may be feeling. Gone is the misgiving that legitimate science has no place for the emotions evoked by seeing a keeper tenderly coax a baby pachyderm: elephant emotions and our own are part of good life and good science.

New knowledge brings new challenges. Once we are aware that Jenny's self-injury is akin to that of a human prisoner, and that the heart-wrenching calls of Lisa for her baby are echoed in the wails of a woman who has just lost her child, what comes next? What do we do with this awareness? The Sheldrick Trust, PAWS, and the Elephant Sanctuary in Tennessee provide examples of how people with this knowledge can help elephants rebuild to a vital future, but their work alone cannot stem the surge of elephants in need, nor are sanctuaries designed to address systemic issues that underlie human-elephant conflict and habitat destruction. Something else is needed if elephant culture is to survive. The task before us, then, is to translate knowledge of trans-species brain, behavior, and psyche into elephant conservation.

It may sound easy to take what we know and put it to work, but as it turns out, the space between what we think and what we do can be quite substantial. First, not all knowledge is the same, nor are its uses. Some learning comes as a surprise, so its ultimate use is less obvious. The microscope opened horizons to an unknown world, and the theories of Einstein started society on an unanticipated path that caused the physicist himself to question the wisdom of his role in discovery. Like its creation, the application of new knowledge involves more than one person.

A single event may be credited for catalyzing one of the philosopher Thomas Kuhn's paradigm shifts, but most often the road to revelation turns out to be paved with a steady accumulation of facts and theories. Even though Friedrich Kekulé happened upon the benzene ring formula in a dream, it was no coincidence that Alfred Wallace and Charles Darwin had the same idea about natural selection. Not discounting genius, both were part of the same nineteenth-century social and intellectual matrix that infused perception and concept. Discovery comes from "particular brains . . . within whole bodies that have been profoundly conditioned from birth by others around them to notice what the group takes to be essential, and to deny or ignore as the group dictates or insinuates."[2] In a more diffuse and indirect way than a parent, culture influences how we think and feel, and what we come to know and do.

But like habits, culture can be slow to change: witness how long animal emotions were denied despite Darwin's theories. Consequently, it is not surprising that the vast amount of knowledge about elephants has not benefited its subjects. Elephants are not treated much differently than they were in the mid-eighteenth century: they are the objects of awe and conservation, yet legally hunted, made captive, abused, and forced to labor for human gain. What then has research and learning served?

According to Thomas Huxley and Johann Wolfgang von Goethe, knowledge obtains meaning only when joined with action. This connection, the link between ideas, meaning, and action, is central to the research of Arthur Glenberg, a psychologist at Arizona State University and the University of Wisconsin, where he is principal investigator of the Laboratory for Embodied Cognition. The lab's motto, *ago ergo cogito*—I act, therefore I think—could have been written by Goethe himself. Glenberg recounts how his research on mirror neurons changed him: "I became a vegetarian [because] it became clear to me as a consequence of these theories of embodied cognition that virtually all animals are thinking, and it is difficult to draw a line between those who are thinking and those who aren't." Later he put even more succinctly: "I avoid meat because I can't find a moral justification for the production of food from thinking and feeling animals."[3]

Statistically, Glenberg is part of a minority. A recent poll found that while 73 of the American public is concerned about the state of the environment, only 10 percent act on their concern.[4] This gap, the space between scientific knowledge and its employment, is central to elephants as we learned in the case of the captive industry. However, the contradictions surrounding elephants extend deep into conservation policy, and South Africa's policy revisions that include the reinstatement of culls—mass killings—provide one good example.[5]

In March 2007 the *New York Times* reported on a new plan by South Africa's environment minister "to control the nation's booming elephant population." The government "contemplates resuming the much-criticized killing of excess animals, but only after thorough scientific study and as a last resort."[6] A little over a year later, in May 2008, South Africa overturned its ten-plus-year moratorium and approved culls as an option for controlling elephant populations. The decision was based on a series of scientific activities that culminated in two documents, the "Summary for Policymakers: Assessment of South African Elephant Management 2007" and its source, the *Assessment of South African Elephant Management 2007*, a several-hundred-page tome written by sixty-two and reviewed by fifty-seven experts. The documents were produced at the behest of a science round table that had been convened to advise the South African minister of the Department of Environmental Affairs and Tourism (DEAT). In a public announcement concerning elephant management, Marthinus van Schalkwyk stated that elephant densities have "risen so much in some southern African countries that there is concern about impacts on the landscapes, viability of other species, and the livelihoods and safety of people."[7] That is, elephants might irreversibly damage ecosystems and thereby threaten biodiversity and present a danger to people, their property, and other species. The *Assessment*'s goal was to "gather, evaluate, and present all the relevant information."[8] In tandem with its companion document, it presents an impressive, comprehensive reference.

But the documents leave the reader mystified for several reasons. The messages conveyed by the *Assessment* are confusingly contradictory, and scientific analysis and emergent policy seem sorely mismatched. A key chapter concludes: "Overall, our assessment is that while the impacts of high elephant concentrations may bring about local changes in vegetation and associated animal species, and hence local biodiversity, this need not be the case at the wider ecosystem level."[9] No known extinctions have resulted from high elephant densities, and unless damage is extreme, the authors say, many plant populations will recover.[10] Elephants are found to enhance ecosystem productivity as part of the delicate chain of plant-animal-soil interactions that have evolved: when large herbivores are excluded, the chain begins to unravel.[11]

Uncertainty effects are difficult to "untangle." Isolating the effects of elephants from other factors that influence an ecosystem, such as fire and extreme variations in the water table, is tricky. How and how fast vegetation will respond to changes effected by elephants are unknown ("very little data"). Furthermore, there is "no recommended density of elephants to manage such changes," and any preferred density would depend on "particular management objectives."[12] In other words, elephants do make a mark on the landscape, but there is little information to ascertain whether it is deleterious or integral to the ecosystem under present conditions. The science is insufficiently accurate to begin on-the-ground action.

Neither did the authors of the *Assessment* find substantive evidence with respect to the second issue of concern, human-elephant conflict: "Elephants and people interact with each other both directly and indirectly, and positively and negatively. Our assessment has shown that most of these interactions occur within conservation areas and are predominantly positive. Further, levels of direct conflict between humans and elephants outside protected areas are generally low." While crop damage and injury do occur elsewhere in Africa, the fact that, in essence, all elephants in South Africa are contained behind fences and "the overwhelming majority of interactions between tourists and elephants in protected areas are positive" means that there are virtually no incidents within the geographical purview of the *Assessment*. The few incidents that do occur are inside the protected areas or in captivity.[13] Human-elephant conflict is negligible.

The *Assessment* and its Summary also conclude that "there are scientific reasons to suggest that elephant [*sic*] have a higher degree of sentience than the vast majority of mammal species" and that "the level of awareness and empathy exhibited by elephant suggest that they might be considered to have a limited form of 'right of privacy.' . . . In elephant there is reasonable cause to suggest . . . suffering includes emotional stress, for instance for fear based on past experience, or through witnessing harm to another elephant." The authors straightforwardly declare, "Knowingly causing unnecessary suffering to any sentient organism is unacceptable and forbidden by law."[14] Nonetheless, despite the recognition that elephants are sentient and susceptible to emotional trauma, and that there was no conclusive evidence showing that elephants in South Africa posed a safety or ecological threat, the new government policy did not withdraw the claim of an "elephant problem," and in fact included "lethal management" to address that "problem." Culls are expected to commence sometime not far in the future: events that are known to cause considerable "emotional stress."

Beyond this blatant contradiction, a "thousand cuts" of conflicting, imprecise, and ambiguous renderings of the information create confusion rather than any decisive, declarative thrust. For example, it is unclear how and for what purpose, or even whether, the science was compiled and used. On one hand, Minister van Schalkwyk insists that "elephant population density . . . has risen so much in some southern African countries." This confident statement suggests that the numbers of which he speaks are derived from research and science-based statistical estimates. Indeed, he lauds such efforts: "I am therefore delighted with the progress that has been made by the elephant science community (as reported earlier by Dr Bob Scholes and Professor Graham Kerley) since I announced in February last year that this Department would contribute [five million rand] to enhance elephant research in South Africa." He goes on to claim, "'The 2007 South African Assessment of Elephant Management' is the first step in a long term Elephant Research Programme that will continue to inform our future law making and management practices. This research is crucial to the continuing debate whether elephant numbers need to be reduced in South Africa, and if so, how."[15]

However, despite the minister's claims, the scientists found no substantive evidence for concern, and they do not appear to agree on the relationship between the *Assessment* and policy. Indeed, they declared that "the Assessment itself does not constitute policy at any level, although it is hoped that it is relevant to the process of policymaking at all levels, from the individual protected area through provincial, local, national, regional and international policy."[16] There is usually an effort to demarcate science and its application, but it is not without precedence that science and policy have become interlocked. A 1994 U.S. Northwest Forest Plan was initiated in the United States during the Clinton administration to resolve economic interests of the timber industry and conservation. After a scientific assessment had been conducted, Forest Plan scientists were charged with developing specific policy options. The elephant *Assessment* makes no such transparent recommendations. The *Assessment* scientists, while charged with an equally significant and politically sensitive issue, seem to have held back, or have been held back, from a more explicit marriage with policy. Yet policy recommendations in the *Assessment* are implicit.

The *Assessment* immediately preceded the decision-making process and contained a detailed map to help decision makers navigate the information. Furthermore, upon closer inspection, woven throughout the intricate complex of data and concept, is a thread of specific policy. At the level of organization, the *Assessment* follows a logical series of chapters describing the history and science of matters pachydermatous—biology, biodiversity, populations, and so on—and concludes that there is no elephant "problem"; then it makes a bewildering leap to a detailed analysis and list of solutions to the (nonexistent) problem. Four of the *Assessment*'s twelve chapters are devoted to elephant control (culls, birth control, further constriction of habitat, repulsion with chili spray).[17]

Scant attention has been devoted to more ecologically sound and scientifically consistent approaches, such as expansion of elephant habitat within a conserved landscape that includes both elephant hot spots and compatible human land use, as suggested by Rudi van Aarde, a South African scientist who participated in the round table and *Assessment*. He has argued for the establishment of "megaparks" and corridors connecting the fragments of conserved elephant habitat: alternative approaches that reflect, rather than contradict, scientific findings. Such designs more closely approximate the historical range and ecology of elephant society in ways that would preempt many problems associated with isolation and the inadequacies of extant park dimensions. Some corridor efforts are under way—for example, a Transfrontier Conservation that joins with Mozambique—but they play a min or role relative to methods of direct control.[18] Van Aarde is not alone. Others, including scientists from the Amboseli Elephant Research Project (AERP) and the Animal Rights Africa spokesperson Michele Pickover have made similar suggestions.[19] But the marginalization of these approaches brings up another inconsistency between what the government says and what is reflected in policy.

While van Schalkwyk states that consultations were extensive, including "detailed interaction with leading elephant scientists," the knowledge and opinions of the Amboseli Elephant Research Project (AERP), comprising acknowledged experts, indeed the premier leaders in the field, who have among them clocked more than two hundred years of detailed study of African elephants, have been ignored or for the most part dismissed. Authors of the AERP *Statement on Elephant Culling* say in no uncertain terms that "it is our considered opinion that killing elephants to reduce local population density ('culling') is unnecessary, unimaginative and inhumane . . . ineffective or retrogressive . . . outdated," and that "culling is, from an ecological point of view, unnecessarily destructive and invariably unjustified and from a social, behavioral and cognitive point of view, unethical."[20] The unequivocal statement, nearly devoid of any qualifiers that would denature the strength of the message, casts serious doubts on whether the South African policy is indeed informed by science. It also raises the question of why the *Assessment* authors find it necessary to gain new knowledge and engage in further research relative to the decision at hand, given what is already known. It is unclear what would supercede what has already been meticulously observed, documented, and analyzed by the expert Kenyan research team that bases its conclusion on more than thirty years' experience, as well as that of the research reflected in the *Assessment*. There are other contradictions between what is documented and proposed and what is concluded and executed.

The minister and the *Assessment* both make a point that extensive consultation was made with a variety of "stakeholders for more than a year before we put pen to paper." However, given what was reported, it appears that only a narrow band of interests and values have been represented: namely, variations on the modern, colonial, European-based worldview which has been absorbed by indigenous cultures.

There appears to have been an almost complete exclusion of African indigenous cultural views and values concerning human-elephant interactions. No chapter includes indigenous knowledge (so-called traditional ecological knowledge [TEK]), and in the chapter on "Interactions Between Elephant and People" only a few sentences acknowledge indigenous views at all. Even the acronym HEC, for human-elephant conflict, fails to distinguish among humans who derive from distinct heritages that radically differ in values of, behavior toward, and relationships with elephants. For example, the *Assessment* acknowledges that "elephants are prominent in African folklore and have particular significance for clans for whom they are totem . . . [and who] value elephants . . . as part of their cultural heritage. . . . However, while communities living adjacent to South African protected areas allude to the cultural value of elephants and the historical relationships between them and people, . . . little has been published or is known about the details of these."[21] It may be true that relatively little has been published, given that indigenous cultural knowledge transmission is usually orally based, and that until the 1990s these societies were politically and legally marginalized and silenced. Colonization brought industrialization and urbanization, which supplanted and disabled traditional economic systems, which in turn led to migration from homelands into the cities and the loss of sacred traditions and land stewardship. Values of the dominant European-based culture undermined the relationship with nature of a significant segment of the population.[22] Further consultations with stakeholders do not require written documentation, or if such documentation were needed, it could be acquired through transcription and translation. It is therefore unclear why there is "little known" about these cultural values and history.

Prehistoric human-elephant interactions are largely described as negative and hostile. "Elephants and people have interacted in Africa for thousands of years"—a statement that seemingly implies a peaceful coexistence, given the numbers of elephants living in proximity to many indigenous cultures when colonists arrived—but the "advent of cultivation . . . probably changed the relationship between the two species from one of 'a mild predator/prey interaction' to one that was 'fundamentally competitive.'"[23] The evidence for such an early "predator-prey" model of interaction is sketchy, referencing only two publications that are described in speculative language. The authors admit in a few sentences that there is indirect evidence of positive interactions, such as elephants banishing the tsetse fly in some areas. Only in one sentence do they note that "various African societies have totem and folklore about elephants indicative of respect." Then, suddenly, the reader is catapulted from prehistory to colonial times with a discussion of present-day values and attitudes about elephants. Not only is early history described largely by conflict, but the absence of serious, in-depth treatment of traditional indigenous views and values that persist today effectively erases the history of the relationships between elephants and all cultures indigenous to South Africa before colonization.[24] The link between the decimation of wildlife and that of indigenous peoples has long been denied or ignored. Lost in the shuffle is the fact that both sets of cultures, elephant and human, were flourishing when colonists arrived and radically declined after European settlement.

Recently the Department of Environmental Affairs and Tourism (DEAT) was strongly criticized by numerous stakeholder groups for sidestepping the formal stakeholder process in the development of "Norms and Standards for Damage-Causing Animals." In a public statement,

Animal Rights Africa spokesperson Michele Pickover cited political and cultural bias in admonishing DEAT on ethical and scientific grounds.

> Apart from the obvious ethical issues, there is no scientific data that supports the case for the use of these cruel lethal methods. Animal Rights Africa is therefore totally baffled by DEAT's current disingenuous stance. We will not allow them to push through an illegitimate policy document. . . . This lobby unashamedly promotes a gratuitous culture of killing and violence and still reflects the rhetoric, language and chauvinistic mindset of wasteful, cruel and inhumane colonialism and apartheid which exploited Africa, its peoples, and its wildlife.[25]

DEAT later reportedly issued an apology for its transgression.

The South African journalist Mike Cadman brings up another factor relating to cultural bias: the strong influence the military tradition exerts on conservation.

> It's important to remember that throughout colonial Africa former soldiers dominated conservation and became "wardens" and senior "officers" in nearly all large colonial game parks and game departments, starting in the late 1890s and persisting for almost one hundred years. Game departments were structured along military lines: mandatory saluting of "officers"—senior staff, usually white males—military-type parades, inspections, and so on.
>
> Many white males, who play a significant role in elephant management in South Africa, either experienced or have close friends or family members who served in the military. The fact that their thinking drove much of modern conservation policy up until about the 1980s is critical. Military thinking, in simple terms, has two components: defense, which is seen as an attempt keep things the same by repelling invaders, and

> attack, which is to try and keep things the same by attacking potential invaders on their turf. Consequently, much of our conservation policy in the past has been based on management by intervention. Soldiers and armies are expected to be aggressive. "If it doesn't look like what we think it should, we'll make it so." Scientists are playing a more constructive role today, but, as recent policy changes indicate, some are still trapped, albeit unconsciously, in the "old" mindset of military strategy. Similarly, some of the new leaders of SANParks [South Africa National Parks] and others in positions of influence, admire the various regional liberation movements which participated in armed resistance against colonial rule, and thus, perhaps also subconsciously, perpetuate the military mindset in wildlife management.[26]

The South African Park experience is not unique; the link between land governance and use and the military is common to other nations.[27]

The *Assessment* presents other puzzlements, substantive inferential gaps that do not fit together logically. Here and there in the documents, beneath well-crafted documentation, there is a subtle undercoat of guarded implication that hints at an inevitable imperative to cull and control. In a chapter that describes the process of decision making, input from science and society is described as filling a "toolbox of available interventions"; among the tools are "culling, contraception, translocation etc."[28] Later, in the Summary, even though neither appreciable HEC nor menacing threat to biodiversity has been found, the authors suddenly imply that these *are* pressing problems that need immediate attention: "Culls and translocation are the only methods to reduce elephant densities (and thus local impacts) in places where intervention is urgent, i.e., taking effect immediately or within five years." The line between the hypothetical and reality have converged. The authors go on to say that once a policy of culling is adopted, it must be "continued indefinitely or until replaced by another method."[29] The implication seems to be that even though scientists did not find evidence for present concern, culls are imperative and will probably be a method of choice in the long term—the reasons for which are unclear, leaving the reader with the impression that the nonexistent elephant problem indeed exists somewhere in South Africa and that culls are the best primary solution.

After an introductory discussion and background, the Summary reviews what is known about elephant effects and acknowledges the limits of that knowledge. Following a section entitled "Will Elephant Numbers Regulate Themselves?" the authors begin by saying, "there is no observational data to answer this question definitively, nor is there likely to be within the next decade."[30] Abruptly, after a brief discussion of whether elephants will regulate their numbers in these unnatural circumstances, attention is again directed to methods of control and the economic benefits of elephants. Almost imperceptibly, we have moved from a state of scientific uncertainty and no imminent problem to an outline of methods of "solving" the problem and an outline of fiscal benefits. Through this shift in focus, the science is lost. Lost is the line of logic laid out in the *Assessment* that would take us from science's recognition of elephant sentience and vulnerability, from the well-being of broad scale ecosystem and human safety to the conclusion that elephant control and culls are not necessary—indeed, that they would, if implemented, break the law by "knowingly causing unnecessary suffering to any sentient organism."[31]

Ironically, the *Assessment* does retain a thread of logic and an appreciation for elephants' humanlike sensibilities even in the design of translocation and killing. The most ethical approach, according to the authors, is to kill entire family groups; in theory, the effect would be similar to that of the more disruptive "disturbance culls," which are intended to compel survivors "to move . . . around the landscape."[32] The term *disturbance* recalls a concept from the early days of the development of ecology as a field. Events such as wildfire and floods, while appearing to damage a forest or savannah ecosystem, were branded as disturbances vital to sustainability because they encouraged the ecosystem and its residents to adapt and evolve. Referring to planned elephant kills as "disturbances" implies that culls are natural events to which the animals can and must adapt.

Rather than leave orphaned young who, like the young bulls, might mature into traumatized survivors with troublesome behavior, the plan would result in whole families being culled—leaving no physical witnesses. But elephant society is not composed of discrete, isolated units arrayed like chess pieces; as we have learned, they are made up of fission-fusion nodes in fluid, highly connected, multitiered social and psychological networks with communication systems capable of bridging miles.[33] It is therefore unlikely that a family cull will go unnoticed and unmourned. In fact the authors do acknowledge the complexity of elephant society and sensibilities, that culls and live removals will cause an "inevitable" "disruption of the complex social network," with "consequences [that] include long-term stress to the population." Shortly after acknowledging these effects, though, the authors cite a series of personal communications attesting to "successful" family group culls in various parks and reserves, with "no major disruptions to the animals that remained behind."[34] There are serious stress effects, then there are not.

The analysis of lethal management is filled with similar inconsistencies and inferences. The authors state, for example, that only one study (cited as a presentation at a scientific meeting that year) has assessed long-term effects of culls. That study found that "elevated stress levels" are evidenced "over a decade later" (which implicitly defines "long-term" as a decade or more). However, data and research, even some from South Africa, are ignored and discounted as evidence of long-term effects: the abnormal behavior of our angry young bulls, which surfaced after approximately a decade as a result of their experience as infants, and that of multiple cases observed in various parts of Africa, all of which are documented not in the less formal and far less scrutinized venue of a conference presentation, but in peer-reviewed, highly regarded scientific journals.[35]

What emerges from both documents is the sense that hard work, fine scholarship, and a wealth of information crafted into elegant text has been unable to escape unscientific reshaping and omission in a patter that strongly suggests that the end product serves not science but predetermined policy.

This brief analysis of the South African *Assessment* and Summary might be considered picayune academic criticism rendered through selective quotations if it was not for the Sherlock Holmesian "curious incident of the dog in the night-time" character of the *Assessment*. Beyond the complex of small to large inconsistencies and internal contradictions, there is a glaring absence in the scientific treatment of the subject and its definition of the (spurious) problem. Behind the meticulous science, the most obvious variable remains cloaked and unrepresented: human behavior. Modern humans' values and behavior are the *causes* of perceived risks to human safety and property and ecosystems and biodiversity, and of elephant behavioral anomalies. In the language of scientists, the neglect of this variable constitutes a "framing error," a problem inadequately or erroneously defined. Even though there is an entire chapter on "Interactions Between People and Elephants," its most salient finding did not make its way into the Summary: modern human concepts and agendas are the *cause* of the so-called and unsubstantiated elephant problem. Efforts are directed toward how to curb the *consequences* of human actions, not to examine how human behavior and beliefs are at cross-purposes with what is supposed to be an agreed-upon objective: conservation of biodiversity via a systems approach. According to one scientist, "the way the elephant issue was framed was about the same as one would use systems theory on a goldfish bowl." Nature's rhythms and patterns do not conform to those created and imposed by Western humans.[36]

The Amboseli elephant researchers headed by Cynthia Moss make the same point:

> Contemporary ecological thinking, supported by a growing body of knowledge, holds that nature including systems with people as actors—is *about* change. Ecosystems are driven by variation in weather, climate, soils and hydrology, and modified by interactions between species. Change occurs at differ-

> ent scales of geography and time, in varying manifestations, over years, decades or millennia. Under this view, biodiversity is maximized by promoting spatial patchiness and temporal variation. The alternative is a futile struggle against natural processes to force nature "to behave."[37]

Modern humans, who constitute the overriding agent of change, were excluded from the *Assessment*'s problem definition. But why has scientific evidence been neatly sidestepped and logic derailed?

Economics is one possible answer. Tourism is a huge economic boon to South Africa, and one might think that the lure of elephants would deflect efforts to limit them. But as the authors of the Summary note, maximum tourism benefit can be obtained with a relatively small number of elephants: "Having more elephants . . . does not imply that the net economic value will increase proportionately."[38] In addition to being visual commodities, elephants can add to the coffers through products made from their bodies. A table in the chapter entitled "Lethal Management" provides estimates of profits and costs associated with culls in Kruger National Park.[39] For an annual cull of eight hundred elephants, the authors calculate a profit equal to about a half-million U.S. dollars per year; if ivory were included, it would "effectively double the profit."[40] Kruger is the largest holding of elephants, but there are other parks and privately owned reserves.

Ivory is no small game. Along with Namibia, Zimbabwe, and Botswana, South Africa participated in a United Nations–sanctioned auction in late 2008 of more than one hundred tons of stockpiled ivory that was to be sold exclusively to Japan and China. The sale was approved by the Convention of Trade of Endangered Species of Wild Flora and Fauna (CITES), an organization established to ensure species survival.[41]

Economic potential under a new policy that permits culls is difficult to ascertain in South Africa because of the past moratorium and laws. However, a comparative study on Botswana, Zimbabwe, and Namibia confirmed that "elephants are . . . quite important as generators of income both nationally and for local communities." Cull by-products and trophy hunting accounted for an estimated 29 to 56 percent of the value of elephants in Botswana in the years 1989, 1990, and 1992, tourism the rest.

Not much different than in the days of Ernest Hemingway, elephants continue to be regarded largely as instruments of commerce. This attitude is reflected by Randall Moore, whom we met earlier and who runs a safari camp as well as the first elephant training center in Africa: "It is evident that Africa's remaining wildlife will have to pay its own way if it is to survive." African elephants have joined their Asian cousins by becoming not only visual targets for the tourist trade but laborers. Moore employs elephants in the tourist off-season by "deploy[ing] his elephants to the rural areas on the Okavango periphery, where they happily haul ploughs to develop fallow land for crops."[42] "Happily," indeed. Reestablished elephants in parks like Pilanesberg draw a steady flow of national and international visitors. The success of such parks, enhanced by the magnetic attraction of elephants for humans, further fuels a demand for both the capture of wild elephants and captive breeding.

South African trade and hunting currently occur on private game farms and in provincial parks, but the "price per elephant, whether as a live sale or for a hunt is very high." The sale of a trained elephant nets the equivalent of between $57,000 and $110,000, and juvenile elephants sell for $5,000–$50,000. A live bull, for which prices depend on tusk weight, costs approximately $10,000 if its tusks weigh more than sixty-five pounds, or $50,000 if used for a hunt.[43] As a "renewable resource," elephants provide a steady, lucrative source of profit through culls, hunts, and tourism.

But economics alone is too simplistic an answer. Not all the money brought in by all the elephants in the world could keep an entire nation afloat. Just as scientists specified that estimates of appropriate elephant densities depend on the objectives specified, economic success is relative to a particular set of objectives and players. Economics and politics are tied together. Steve Best, an ethicist at the University of Texas, speaks about the parallels between human and species apartheid in South Africa:

> Apartheid was a brutal system of class and racial domination maintained by repression, violence, and terror, whereby a minority of wealthy and powerful white elites exploited and ruled over the black majority. . . . Under the pseudo-progressive guise of progress, rights, democracy, and equality, leftists, communists, democratic humanists, black nationalists, and community activists murder animals no different than white, racist, Western, capitalist, imperialists. Consider, for instance, the Zimbabwe "Campfire Conservation Association" that lobbies the U.S. Congress for funds to kill elephants for community benefit. Through a blatant discourse of objectification, Campfire member Stephen Kasere unashamedly reveals his speciesist outlook: "We just want the elephant to be an economic commodity that can sustain itself because of the return it generates. Ivory is a product that should be treated like any other product."[44]

Best and others point out that the issue is not conflict between species but rather the imposition of one political agenda meeting resistance from a culture required to live as economic products.

Are elephants going extinct? Some say the glass is half empty. Many conservationists maintain that lucrative ivory and commodity sales will escalate the already increased illegal poaching. According to one prediction, local elephant populations could be extinct within a decade. Not only are poaching-related elephant death rates higher than during the 1980s, when an international ivory trade ban was in place, but elephant populations, one million as recently as twenty years ago, have shrunk by half.[45] In the state of Karnataka, India, more than seven hundred elephants have died since 2003—about one death every three days.[46] Critics say that estimates of elephant populations are famously unreliable and that the animals are not faring as poorly in the wild as some have proclaimed. The International Union for the Conservation of Nature (IUCN) 2008 Red List of Threatened Species noted that "the oft-repeated global population 'estimate' of about 40,000 to 50,000 Asian elephants is no more than a crude guess, which has been accepted unchanged for a quarter of a century. . . . With very few exceptions all we really know about the status of Asian elephants is the location of some (probably most) populations, with in some cases a crude idea of relative abundance; and for some large parts of the species range we do not even know where the populations are, or indeed if they are still extant." While the report's authors felt that the "overall population trend of the Asian elephant has been downwards, probably for centuries," and that "this remains the case in most parts of its range, but especially in most of the countries of South-east Asia," they also noted that "within India, there is evidence that the large population in the Western Ghats in the south of the country has been increasing in recent years due to improved conservation effectiveness."[47]

The IUCN Red List reduced the threat level for African elephants from "Vulnerable" in the 2004 Red List to "Near Threatened" in 2008: "The change in status reflects recent and ongoing population increases in major populations in Southern and Eastern Africa. These increases are of sufficient magnitude to outweigh any decreases that may be taking place elsewhere."[48] Keith Lindsay of the Amboseli Elephant Research Project stresses that while current elephant numbers suggest elephants are not on the brink of extinction, more attention should be paid to trends in such critical factors as habitat quality and quantity, as well as human land use in areas they share with elephants.[49] Indeed, numbers can tell us only so much. History has shown that populations thought to be robust have crashed precipitously in the space of only a few years, a dynamic that may resemble a "black swan problem," where a calamitous event is largely unpredictable and there are unanticipated consequences.[50]

Even if elephant numbers don't imply a threat to the species, daily dispatches provide evidence that their living conditions are increasingly stressful. One report from India illustrates their plight: "With the increasing . . . death rate of elephants in Karnataka state, the rare sight of pachyderm strolling in the wild may even cease to exist. The elephant's survival is at risk as they are poached for their valuable ivory tusks or are electrocuted by hassled farmers to protect their crops."[51] From the victims' point of view, from the perspective of elephants, statistics merely objectify an experience of progressive genocide approaching a fixed number of survivors to be held and perpetuated in captivity or used to service the demand for the thrill of hunting and the commodities made from their bodies. Indeed, many fear that African wildlife are on their way to becoming much like domesticated stock through the burgeoning business of game ranching. There are even efforts afoot to move "game ranching" from the purview of the Department of the Environment to that of agriculture, thus converting wild animals into "production units."[52]

The science historian Jane Carruthers has delved into the ideological roots and evolution of South African attitudes toward wildlife. She makes the point that even linguistically, the line between domesticated animals and free-ranging ones is blurred. The Afrikaans language does not distinguish between "wildlife" and "game," using the term *wild* (pronounced "vilt") for both: neither is there a distinction between "ranch" and "farm." While ranching in South Africa refers to the animals living in a semiwild state, where certain ecological factors are considered and only minimal sustenance is provided beyond what is naturally available (as opposed to regular feeding on farms), the objective of both enterprises is the production of meat and animal products. There are five thousand game ranches in South Africa and more than four thousand with mixed game and livestock, covering some 13 percent of the land, compared with 5.8 percent for all officially declared conservation areas, of which the national parks account for only about half.[53] The population of South African elephants comprises approximately thirty-four individual groups, with a thirty-fifth under the official designation "Private reserves" in the Western Cape, Eastern Cape, KwaZulu-Natal, Mpumalanga, Limpopo and North West provinces. Of these thirty-five populations, twelve are in national parks, on state lands, or in provincial nature reserves, and the rest are located on private land.[54]

Understandably, the barrage of statistics and conflicting scientific opinions makes elephant conservation confusing. But confusion comes not only from the numbers or the disagreement among scientists: after all, science is built on the premise that the quality of knowledge profits from scrutiny and comparison of differing theories and evidence. The real contradiction lies within, not between, individual scientists. The same individual can on one hand laud elephants, speak of them in terms we use for ourselves, then turn to the need to "manage" them, advocating measures up to and including systematic killing. We have arrived at a threshold suspended between two paradigms, two worldviews, with science as the shared medium forcing a shift and reconciliation of contradictions.

After the conservation savant Richard Leakey visited Kruger National Park and the huge abattoir complex that processed thousands of culled animals, he expressed his alarm and dismay that Africa's wild had been converted into an industrialized farm, and declared that if this trend continued, then wildlife's days "were numbered."[55] Yet commenting on the South African culling decision, close on the heels of Kruger abattoir renovations, he said: "By 1990, long-term research in Kenya and elsewhere had revealed that elephants have highly organized societies and have a surprisingly well developed ability to communicate. We consider them sentient creatures like whales and apes that deserve special consideration when it comes to their management. . . . While I will never 'like' the idea of elephant culling, I do accept that . . . reducing elephant populations may therefore, be a necessary part of population management, and this will be done in a humane and considered manner."[56] It cannot escape the reader that humans are, after all, also apes.

The International Fund for Animal Welfare (IFAW) has spoken out against culls, but other prominent conservation organizations have not.[57] Rob Little of the conservation organization World Wildlife Fund (WWF) typifies the spirit of compromise: "No wildlife manager likes to cull any species. . . . But if any one of the species starts to dominate the food resource of other species or threaten the vegetation of an area, then something has to be done. We believe that the culling . . . should be available as an option."[58] Zoos, too, win support from some members of the conservation community, despite the well-documented ravages of captivity. The veteran conservationist Iain Douglas-Hamilton says that zoos "play a significant role in conservation by stimulating the interest of children and adults."[59]

If the dissonance and ethical double standard sound familiar, it is no wonder. Even though professionals who work with elephants come from a variety of nations, the concepts and methods underlying elephant conservation and science are heavily influenced by a cultural worldview that subordinates elephants, and all animals, to human whims. Science and policy in South Africa are not alone in their contradictions. Modern culture and education are defined by the splintering and parsing of the world into disciplinary, perceptual, and conceptual fragments: mind/body, human/animals, objective/subjective, professional/private. Zoos, circuses, conservation organizations, and science are progeny of Descartes. By articulating a conceptual paradigm of parallel but separately functioning worlds, Descartes and his successors created the foundation of a culture that fosters psychological doubling as a collectively viable method of survival.

Dietrich von Haugwitz provides a more personal view of Cartesian splitting or doubling. He was born on July 17, 1928, in Silesia, eastern Germany, now part of Poland. Despite efforts by his aristocratic family (including the orchestration of an unnecessary appendectomy), the young von Haugwitz was drafted into the German Wehrmacht in 1944, only to be part of the surrender shortly thereafter. Poor and disenchanted with Germany, he eventually immigrated to the United States.

He credits three events with changing him from an "animal lover" to an animal advocate. First was witnessing a bullfight in Mexico, and second was watching a film depicting animal suffering resulting from hunting and vivisection and in slaughterhouses. Von Haugwitz later reflected: "On an emotional level, all of this is terrible, agonizing, almost unbearable, if one is compassionate—and I'm proud to be compassionate rather than unfeeling. However, feeling is one thing, but reason is quite another. And reason tells me that this is the way it is, and it must be."[60] Only when he encountered "the rational argument that validated [his] feelings" in the form of the philosopher Tom Regan's book *The Case for Animal Rights* did von Haugwitz change his mind. Indeed, his rational mind was "the only level on which I was reachable."[61]

A similar mechanism of separation is found in science's insistence that subjective and emotional information be excised from professional observation and analysis. Standards may have changed somewhat since Jane Goodall's early studies, when she was chastised for distinguishing among her chimpanzee subjects with names instead of numbers, but empathy for one's scientific subjects remains cause for suspicion.

A former colleague of B. F. Skinner, the clinical psychologist John Gluck, Jr., made a similar observation about the behaviorist's attitude toward his research. In reminiscing about a former student, Skinner reflected on their work together. Skinner "especially mentioned that they had invented an apparatus that delivered electric shock to rats in learning experiments that was 'fool proof.' He went on to say that neither of them really liked to shock rats so he expected that they had missed out on many important discoveries. In other words, he lamented his sensitivity to delivering painful shocks to sentient beings. The sensitivity, he felt, was irrelevant to his ethical life."[62] Empathy and emotion were scientific liabilities. The Harvard psychologist John Mack writes in exasperation at science's perceptual myopia: "Academic psychology, embodying now a reverence of numbers, tight reasoning, and linear thinking as opposed to intuition, direct knowing, and subjective experience, is likely to look askance at efforts to reinfuse its body with imprecise notions of spirituality and philosophy, from which it has so vigorously and proudly struggled to free itself in an effort to be granted scientific status in our universities, laboratories, and consulting rooms."[63] Compassion that researchers may feel toward their subject matter is not admissible as valid "data," and those who exhibit such emotions may be accused of a lack of objectivity and hence questionable science.

Today this perspective is starting to be challenged, as illustrated in the work of the neuropsychologist Allan Schore, the affective neuroscientist Richard Davidson, and a growing number of other researchers who assert that emotions are integral to how the brain works, decisions are made, and actions are informed. Furthermore, subjectivity, both of the observer and the observed, is no longer considered "polluting" to truth and fact. Nonetheless, science and scientists still seem to exhibit the internal conflicts embodied in psychological splitting. Another personal story may suggest why.

When news reached the world at large that South Africa was about to allow elephant culls, a researcher was approached to cosign a letter of protest on the basis that the science did not justify this decision. While privately supporting the letter and the science cited, however, the researcher would not say so in public; although he was tenured, well established and respected, and financially comfortable, he declined to sign the letter. Someone outside academic circles might find the unwillingness to express one's true scientific opinion perplexing. After all, absent a financial or professional risk, what was there to fear?

The answers offered by historians and other scholars reveal that a complex of factors determines scientists' behavior. One factor is scientists' perception of their role in society and the academy, and of the expectations about their actions. Notwithstanding C. P. Snow's image of scientists as "free-spirits" unhindered by political concerns, David Joravsky, historian emeritus at Northwestern University, argues that "an overwhelming majority of scientists have shown themselves to be single-minded devotees of winning, to the extent that they are venturesome, or of keeping their comfortable jobs, to the extent that they are cautious. They are prizefighters or pay-rollers, or some mixture of the two types." In reflecting on the choices made by physicists involved in making the atom bomb, Albert Einstein criticized what he saw as an intrinsic flaw, the "connection between 'the mechanized and specialized thinking' characteristic of natural science and the stunting of ethical feeling." Joravsky goes farther: "I would turn that harsh judgment into a question. I wonder whether the ethics required for scientific work as for other specialized roles in modern society, the rules imposed by one's 'discipline,' may not be at odds with a broader human ethic."[64] In a somewhat similar vein, Robert Jay Lifton speaks of the Nazi doctors' "Faustian bargain,"—a pact in which one's soul is given in exchange for omniscience.[65] Knowledge, the sense of omnipotence it brings, and the often attendant fame and monetary gain are lures for the scholar in legend and life, particularly with the rise of what the University of California, Berkeley, anthropologist Laura Nader has described as science's corporatization.

Joravsky believes that an ethical breach is "implicit in the scientists' unctuous separation of power over nature, which they seek, from power over people, which they abjure. Modern ideologies, such as nationalism, democracy, and socialism, have intensified and obscured the problem by fusing the individual with the collective, . . . thereby blurring the sense of individual responsibility while increasing the individual's willingness to enhance the collective's power." He concludes that "we are held to our specialized roles by ideologies and paychecks rather than by biochemicals" that bind other animal societies such as ants.[66]

Joravsky's image is evocative of C. G. Jung's description of the intrinsic ethical tear that occurs when an individual subordinates personal responsibility for membership in the group. When an individual scientist becomes defined by her collective role, scientific opinion is silenced by a fear of social rejection and loss of fiscal and intellectual privileges. As we have seen, zoos, circuses, governments, and conservation organizations supporting culls make funds available for elephant researchers. Speaking against captivity and culls based on ecological and other scientific arguments translates to criticism of the institutions that count on these methods of elephant control and economics. In the end, ethical autonomy is weighed against the cost of losing group membership and financial gain.

Another factor is at work, something intrinsic to a collective. Social institutions, whether they be families, communities, or cultures, are innately conservative. They serve as a counterpoise to change, and science, by force of its discoveries of human-animal comparability, is disrupting a culture based on human dominance. The ferociousness with which this self-appointed privilege is defended surfaces subtly in the Summary. After extolling the virtues and humanlike capacities of elephants to the point of a somewhat reluctant admission that the animals may have a right to privacy, the authors of the *Assessment* go on to say that elephants' "capacity for self-consciousness, empathy for other elephants, and problem-solving ability, is on the basis of available information" judged to be "very much lower than that of humans," and "it is humans who define the framework."[67] What this simple phrase suggests is that no matter what science has discovered about human-elephant comparability, nothing is permitted to upset human privilege.

Here we have arrived at the essence within the swirling confusion of fractured facts and quiet messages. Psychological doubling is masked as Cartesian logic not just within the confines of a death camp but, as Lifton predicted, in pervasive and defining cultural norms of perceptions and behavior. Culturally ingrained doubling has permitted the simultaneous pursuit of scientific knowledge in the cause of conservation (humane self) and membership in a social collective sustained by the commodification of elephants (Auschwitz self). Doubling is not limited to scientists, zoos, and circuses. We as the consumers and members of the broader society are also participants.

As the body for which science ostensibly labors, modern culture feeds ethical doubling by encouraging its members to live in ways that keep them separated psychologically from nature. The relationship between elephants and humans is rendered into an adversarial dichotomy with the invention of "human-elephant conflict," dividing a relationship into two parts: a sympathetic elephant "bad guy" and an erring but well-intentioned human "good guy." Television and other media broadcast successions of ecological horrors and animal tragedies, and while our built-in empathy is stimulated by the activation of mirror neurons upon seeing a gorilla lying murdered and an elephant with ivory hacked from his face, these emotions are quickly contained and curtailed by one of several mechanisms: a quick reality-changing switch of channels, an interrupting advertisement for deodorant, or the fearful press to pay the bills, keep a job, keep the peace, and make it through the week. Emotional information that could be used to change beliefs and catalyze action is blocked or diverted; impulses emerge instead as affective release, dissociation, splitting, and numbing to dampen the profound empathetic connection we feel with our animal kin. To feel too much endangers the doubling partition. To identify with the dying glory of a silverback gorilla, the crumpling of the powerful elephant, heralds our own mortality and makes us realize our vulnerability to nature, which pays no heed to human attempts to hide behind a taxonomic curtain. To recognize ourselves in animals' eyes pulls back the curtain exposing the wizard as only a timid fearful soul with no place to hide.

We have discovered why elephant breakdown may be so disturbing.

By coming so close in mind and presence, elephants shake the defensive intrapsychic divide that makes possible doubling and life in twin. The breach of Adam's Wall demands not only a reimagination of human identity but the relinquishment of the psychological, and hence, practical, mechanisms and abilities considered necessary for survival in modern society—even if it has meant the destruction of the charismatic and cherished elephant.

The theme should be familiar. A glance across time and space reveals doubling elsewhere. Just as images of the elephant oscillate between mighty god and circus clown, pictures of the American Indian toggle between contradictory images: light and dark, good and evil, savagery and munificence, damnation and salvation, desolation and paradise, freedom and subordination. Americans were "of two minds about Indians."[68] Just as some scholars have debated current elephant population figures, others wrangle about how many Indians there really were at contact: the "expansion and shrinkage of Indian population estimates correlate with changing attitudes about Native Americans' rights and prospects."[69] So, too, the elephant.

The historian Susan Scheckel writes, "Indian policy called into question the very principles on which the idea of America was founded, threatening to make explicit the contradictions implicit in American ideology and social experience . . . and reveal the deep tensions in the discourses of American nationalism."[70] So, too, the elephant. The Indian "belonged in the American past and was socially and morally significant only as part of that past."[71] So, too, the elephant, banished to reserves, zoo exhibits, conservation areas, celluloid, and symbol. The cowboy movie becomes a safari ride, the Indian reservation, a park.

Most lands stand vacant of their indigenous people. Now it is the elephant who stands defiant, desperately holding onto what remains in hopes that the tide can be stemmed. Now it is the elephant who questions human identity and will. Dare we put into action what we know? Can we reach out, like the elephants have done, and start anew as their brethren?

Epilogue
Quilt Making

Elephant and human societies are linked in prehistory, the present, and with greatest hope, the future. The stories of these elephants are not only what we have done to them but also what has happened to our collective soul. These moments in the present here and now are the ripest moments for a new paradigm with newfound understanding. Like elephants, we humans have watched our families imprisoned and murdered, our children shed blood in our backyards and on foreign turf. We now fight back with a formidable strength that comes from common bonds and love. Elephants and humans have missed each other terribly—we all cry over the bleached bones. Our tears become our resolve.[1]

Humanity stands at the point where the pendulum hesitates in its arc, balanced between two choices. If we fail to heed what the elephants have

told us, fail to act on what we know, then we will lose them, and more. What science has explained about the operations of elephant brains, their joys, their grief, their ways of thinking and psychological vulnerability to human violence holds true for virtually every other wildlife species: parrots, lions, chimpanzees, macaques, whales, cougars, bison, bears, and, it is beginning to appear, fish, reptiles, and invertebrate relatives of the octopus.

On the other hand, if we choose to act on our knowledge of the transspecies mind, we reenter nature and join the animal kin from whom we have become so estranged. Our reunion is possible only if humanity is willing to change. Change brings gain and loss. Like the Little Mermaid, who forwent eternal life in the sea to live with her prince on land, we must relinquish something. To save the elephants, we must let go of the very things that have protected us from being treated like them: our self-appointed dominion and privilege.

But do we have a choice? We have also learned that what befalls the elephants befalls humanity. If we lose the elephants, we lose ourselves. This message is found in theories of the brain and mind and in the nightmares of soldiers, zookeepers, and orphaned elephants, and none are easy to forget. Trauma refuses to stay with the victim alone: "The tissues of community can be damaged in much the same way as the tissues of mind and body, but even when that does not happen, traumatic wounds inflicted on individuals can combine to create a mood, an ethos—a group culture, almost—that is different from (and more than) the sum of the private wounds."[2] This knowledge leads to ethical change: "We are responsible for the other's response. . . . This infinite responsibility entails the imperative to question ourselves and constantly engage in self-critical hermeneutics. It is this critical interpretation that also gives meaning to our lives."[3] We are responsible for the elephants' response.

The path to change is not easy. Becoming conscious of the shadow, Jung writes, cannot be done "without considerable moral effort."[4] The prospect can be terrifying. We have seen what happened to the SS doctor Eduard Wirths when his two worlds came together: the divider collapsed, and he, from what we can gather from records and Lifton's research, imploded. History has borne witness to the fates of others who dared to rattle the gates of the status quo, what T. E. Lawrence called those "dangerous men," those "dreamers of the day . . . [who] act their dream with open eyes, to make it possible."[5] Galileo died under house arrest, Steven Biko was assassinated, Lawrence was driven into self-imposed obscurity, and Timothy Treadwell was pathologized and ridiculed before being consumed by his beloved bears.

But change has already started. It can even be found within the pages of the South African *Assessment:* "In the past 50 years humans have investigated elephants in depth. As a result our knowledge about the ecology, physiology, behaviour, social structure, communication, and mental characteristics of elephants has deepened our perception and understanding of elephants. . . . We therefore need a new ethic as our newly acquired knowledge about elephants requires a redefinition of our relationship with animals that are more like us than we previously realised."[6] New scientific knowledge calls for new ethical practices.

And there is a pantheon of role models to emulate: Dame Daphne, her daughter Angela, and the legion of Elephant Keepers who stitch back torn young elephant lives into vital communities. Pat Derby and Carol Buckley have stepped in to honor and care for elephant elders left stranded from their homelands. Patti Ragan, Gloria Grow, and Jane Dewar rebuild the lives of great apes; Phoebe Greene Linden, Eileen McCarthy, and pattrice jones tend to the shattered psyches of avians; Karen Paolillo holds a piece of Eden for Zimbabwan hippopotamuses; Charlie Russell and Charlie Vandergaw, who live astride the human-bear worlds; Toni Frohoff, Cathy Kinsman, and Robin Lindsay reach out across the waters to cetaceans and pinnipeds; and then there are the Tennessee elephant Divas themselves—hard at work re-creating their own immigrant culture. All are part of a long legacy of human "quilt makers"—those individuals who pick up the pieces left from social breakdown and stitch communities back together.

In recovery from the chaos of slavery, African Americans have nursed, reared, and brought into the world their children and the children of others. Their African ancestors, like elephants, were cull victims, targets of massive violence, whose survivors were translocated to strange lands to live in broken herds bereft of elder leadership. Arriving in North America, African peoples carried with them their values and traditions of attachment, replacing genetically related family members lost in the chaos with other constellations of mothers to rear the children, "othermothers." In the face of social and ecological breakdown, othermothers form a survival strategy adapted to circumstance: "African and African American communities . . . have recognized that vesting one person with full responsibility for mothering a child may not be wise or possible. As a result othermothers, women who assist bloodmothers by sharing mothering responsibilities, traditionally, have been central to the institution of Black motherhood."[7] The bonds that are stitched together "transcend the personal relationship between a mother and a daughter. A nurturing female community of grandmothers, aunts, and friends encircles their 'daughters' in order to ensure some familiarity in their journey into a world characterized by uncertainty and even hostility."[8] Men can nurture, too—the multitudes of men like Ron Grimes and the Kenyan Elephant Keepers who have stepped in to mentor the fatherless young.

Quilt making, resonant with elephant allomothering traditions, creates the possibility of a new, integrated trans-species identity, one in which we each share something of another species' culture. We are able to harness the volumes of scientific knowledge with the wisdom of indigenes, elephants, and other animals to help re-create human culture. Once the shock of trauma sinks in, the possibility for renewal begins.

We need no new research or knowledge before we act. Without immediate and committed action, there will be no mysteries to study, only an empty mirror staring back. The elephants have called. It is time that we join them.

Appendix

Ten Things You Can Do to Help Elephants

1. Learn how to think and live a trans-species life—how you might integrate what you admire in elephants and other animals into your values and lifestyle.

2. Helping other animals helps elephants. Extend the "footprint of compassion" to other species. For example, study your beliefs and behavior and reflect on how they affect animals. If what you do and how you live harms animals, follow your heart and change so that animals do not suffer.

3. Support wildlife protection. Learn how to share the environment with other species so that they are able to live without fear of human harm and can raise their young in healthful, ample habitat.

4. Learn more about elephants. There are many informative resources on the Web; here are just a few:

ELEPHANTS IN THE WILD

The David Sheldrick Wildlife Trust, www.sheldrickwildlifetrust.org
The Owens Foundation, www.owens-foundation.org
Amboseli Trust for Elephants, www.elephanttrust.org
Elephant Voices, www.elephantvoices.org
Elephant Listening Project, http://www.birds.cornell.edu/brp/elephant

ELEPHANT SANCTUARIES IN THE UNITED STATES AND ABROAD

PAWS (Performing Animal Welfare Society), www.pawsweb.org
The Elephant Sanctuary in Tennessee, www.elephants.com
Compassion Unlimited Plus Action Bangalore, www.cupabangalore.org/
SanWild, www.sanwild.org/

5. Support organizations that advocate for elephants and other wildlife and protect their habitats in the wild:

The Kerulos Center, www.kerulos.org
Animal Rights Africa, www.animalrightsafrica.org
In Defense of Animals, www.savezooelephants.com

6. Support local campaigns to move elephants and other animals in captivity to sanctuaries that care for animals in recovery from the trauma of confined captivity. Urge your elected representatives to pass town or city ordinances or statewide laws to ban circuses and other commercial use of wildlife.

7. Do not patronize institutions of captivity: zoos, safari rides, roadside shows, private collections, and events and businesses that use animals for media, entertainment, and profit. Instead, support individuals and organizations that advocate for and protect wildlife cultures.

8. When your local or national newspaper or television station covers elephants or other wildlife, write letters to the editor or station manager. If the media produce informative stories on elephant issues, thank them. If the pieces do not discuss or address the suffering of animals in captivity, write a letter about why elephants and other animals should not be used for entertainment or profit.

9. Share your knowledge. Talk with your friends and family about elephants, and about preserving them in the wild. If you are a teacher, a parent with school-age children, or just someone interested in humane education, find out about sharing your knowledge with class groups and other organizations. Speak out, speak your heart.

10. Buying choices can have a major impact on the habitats of elephants and other wildlife. When giving a gift to a friend, family member, or associate, send a donation in her name to a nonprofit organization that supports wildlife protection. You can also make a big difference just by being a conscious consumer. Look out in particular for:

- Animal products. Don't buy ivory, even ivory labeled "legal" or "preban." Don't buy or use other products that are made from animals.
- Products containing palm oil. Because of increasing world demand for palm oil, habitat land is being taken over at an alarming rate for oil palm plantations. Many everyday products, such as soap, cosmetics, and baked goods, contain palm oil, so read the labels before you buy. Learn more at ran.org/the_problem_with_palm_oil/the_problem/.

- Coffee and chocolate. Increasing demand is leading to loss of rain forest canopy as land is cleared for these crops. Buy only shade-grown and organic products, cultivated without eliminating habitats. Learn more at www.ineedcoffee.com/08/elephant-conflict/.
- Timber and forest products. Look for lumber and other products that have been certified by the Forest Stewardship Council (FSC), an international organization that encourages responsible management of the world's forests. Learn more at www.fsc.org.

Notes

1. The Existential Elephant

1. Asian Elephant North American Regional Studbook, http://www.elephanttag.org/Professional/AsianElephantStudbook.pdf (2005), 47, no. 208.

2. Tilo Kircher and Anthony David, eds., *The Self in Neuroscience and Psychiatry* (Cambridge: Cambridge University Press, 2003).

3. National Museum of Science and Technology, Stockholm, http://www.tekniskamuseet.se/templates/Page.aspx?id=12447.

4. Gordon G. Gallup Jr., "Chimpanzees: Self Recognition," *Science* 167 (1970): 86–87; Gordon G. Gallup Jr., "Mirror-Image Stimulation," *Psychological Bulletin* 70 (1968): 782–93.

5. Helmut Prior, Ariane Schwarz, and Onur Güntürkün, "Mirror-Induced Behavior in the Magpie (*Pica pica*): Evidence of Self-Recognition," *PLoS Biology* 6 (2008): e202; Joshua M. Plotnik, Frans de Waal, and Diana Reiss, "Self-Recognition in an Asian Elephant," *Proceedings of the National Academy of Science* 103 (2006): 17053–57.

6. Marc Bekoff and Paul W. Sherman, "Reflections on Animal Selves," *Trends in Ecology and Evolution* 19 (2004): 176–80; William James, *Essays on Faith and Morals* (Cleveland: Meridian, 1890).

7. Prior, Schwarz, and Güntürkün, "Mirror-Induced Behavior"; Thomas Suddendorf and Emma Collier-Baker, "The Evolution of Primate Visual Self-Recognition: Evidence of Absence in Lesser Apes, *Proceeding of the Royal Society B*, early online publication, February 25, 2009, http://rspb.royalsocietypublishing.org/content/early/2009/02/21/rspb.2008.1754.abstract.

8. K. Carlstead and D. Sherpherdson, "Effects of Environmental Enrichment on Reproduction," *Zoo Biology* 13 (1994): 447–59.

9. G. A. Bradshaw and Robert M. Sapolsky, "Mirror, Mirror," *American Scientist* 94 (2006): 487–89; G. A. Bradshaw and Barbara L. Finlay, "Natural Symmetry," *Nature* 435 (2005): 149.

10. Karyl B. Swartz and Sian Evans, "Anthropomorphism, Anecdotes, and Mirrors," in *Anthropomorphism, Anecdotes, and Animals*, ed. Robert W. Mitchell, Nicholas

S. Thompson, and H. Lyn Miles (Albany: State University of New York Press, 1996), 310; Diana Reiss and Lori Marino, "Mirror Self-Recognition in the Bottlenose Dolphin: A Case of Cognitive Convergence," *PNAS* 98 (2001): 5937–42.

11. On the inclusion of invertebrates see Nigel R. Franks and Tom Richardson, "Teaching in Tandem-Running Ants," *Nature* 439 (2006): 153.

12. Ibid.; Georg Northoff and Jaak Panksepp, "The Trans-Species Concept of Self and the Subcortical-Cortical Midline System," *Trends in cognitive sciences* 12 (2008): 259–64; G. A. Bradshaw and Allan N. Schore, "How Elephants Are Opening Doors: Developmental Neuroethology, Attachment, and Social Context," *Ethology* 113 (2007): 426–36; Benjamin L. Hart, Lynette A. Hart, and Noa Pinter-Wollman, "Large Brains and Cognition: Where Do Elephants Fit In?" *Neuroscience and Biobehavioral Reviews* 32 (2008): 86–98; Jeheskel Shoshani, William J. Kuptsky, and Gary H. Marchant, "Elephant Brain, Part I: Gross Morphology, Functions, Comparative Anatomy, and Evolution," *Brain Research Bulletin* 70 (2006): 124–57; Luke Rendell and Hal Whitehead, "Culture in Whales and Dolphins," *Behavioral and Brain Sciences* 24 (2001): 309–24; Michael Krützen, Janet Mann, Michael R. Heithaus, Richard C. Connor, Lars Bejder, and William B. Sherwin, "Cultural Transmission of Tool Use in Bottlenose Dolphins," *Proceedings of the National Academy of Science* 102 (2005): 8939–43; Victoria A. Braithwaite and Felicity A. Huntingford, "Fish and Welfare: Do Fish Have the Capacity for Pain Perception and Suffering?" *Animal Welfare* 13 (2004): S87–S92; Victoria A. Braithwaite and Philip Boulcott, "Can Fish Suffer?" in *Fish Welfare,* ed. Edward J. Branson (London: Blackwell, 2006), 78–92; Martin Brüne, Ute Brüne-Cohrs, William. C. McGrew, and Signe Preuschoft, "Psychopathology in Great Apes: Concepts, Treatment Options, and Possible Homologies to Human Psychiatric Disorders," *Neuroscience and Biobehavioral Reviews* 30 (2006): 1246–59; Gordon M. Burghardt, *The Genesis of Animal Play: Testing the Limits* (Cambridge: MIT Press, 2005); Gordon M. Burghardt, Brian Ward, and Roger Rosscoe, "Problem of Reptile Play: Environmental Enrichment and Play Behavior in a Captive Nile Soft-Shelled Turtle, *Trionyx triunguis,*" *Zoo Biology* 15 (1998): 223–38; Gordon M. Burghardt, "Precocity, Play, and the Ectotherm-Endotherm Transition: Profound Reorganization or Superficial Adaptation?" in *Handbook of Behavioral Neurobiology,* vol. 9, ed. Elliott. M. Blass (New York: Plenum, 1988), 107–48; Jessica F. Cantlon and Elizabeth M. Brannon, "Shared System for Ordering Small and Large Numbers in Monkeys and Humans, *Psychological Science* 17 (2006): 401–6; Jackie Chappell and Alex Kacelnik, "Tool Selectivity in a Non-Mammal, the New Caledonian Crow (*Corvus Moneduloides*)," *Animal Cognition* 5 (2002): 71–78; Jackie Chappell and Alex Kacelnik, "Selection of Tool Diameter by New Caledonian Crows, *Corvus Moneduloides,*" *Animal Cognition* 7 (2004): 121–27; Kristopher P. Chandroo, Stephanie Yue, and Richard D. Moccia, "An Evaluation of Current Perspectives on Consciousness and Pain in Fishes," *Fish and Fisheries* 5, no. 4 (2004): 281–95; Kristopher P. Chandroo, Ian J. H.

Duncan, and Richard D. Moccia, "Can Fish Suffer? Perspectives on Sentience, Pain, Fear, and Stress," *Applied Animal Behavior Science* 86 (2004): 225–50; Lily S. Chervova, "Pain Sensitivity and Behavior of Fishes," *Journal of Ichthyology* 37 (1997): 98–102; N. S. Clayton, D. P. Griffiths, N. J. Emery, and A. Dickinson, "Elements of Episodic-Like Memory in Animals: Philosophical Transactions of the Royal Society of London," *Biological Sciences* 356 (2001): 1483–91; Ana P. daCosta, Andrea E. Leigh, Mei-See Man, and Keith M. Kendrick, "Face Pictures Reduce Behavioural, Autonomic, Endocrine, and Neural Indices of Stress and Fear in Sheep," *Proceedings of the Royal Society of London*, series B, 271 (2004): 2077–84; Marc Hauser, "Our Chimpanzee Mind," *Nature* 437 (2005): 60–63; Lee A. Dugatkin, "Animal Behaviour: Trust in Fish," *Nature* 441 (2006): 937–38; Samuel D. Gosling and Oliver P. John, "Personality Dimensions in Non-Human Animals: A Cross-Species Review," *Current Directions in Psychological Science* 8 (1999): 69–75; Dale J. Langford et al., "Social Modulation of Pain as Evidence for Empathy in Mice," *Science* 312 (2006): 1967–70; C. A. Foley, S. Papageorge, and S. K. Wasser, "Noninvasive Stress and Reproductive Measures of Social and Ecological Pressures in Free-Ranging African Elephants," *Conservation Biology* 15, no. 4 (2001): 1134–42; Kathleen S. Gobush, B. M. Mutayoba, and S. K. Wasser, "Long-Term Impacts of Poaching on Relatedness, Stress Physiology, and Reproductive Output of Adult Female African Elephants," *Conservation Biology* 22 (2008): 1590–99.

13. Gordon G. Gallup Jr., "Self-Recognition in Primates: A Comparative Approach to the Bidirectional Properties of Consciousness," *American Psychologist* 32 (1977): 329–38.

14. German E. Berrios and Ivana S. Markova, "The Self and Psychiatry: A Conceptual History," in Kircher and David, *The Self in Neuroscience and Psychiatry*, 9–39.

15. William James, *The Principles of Psychology* (1890; New York: Dover, 1950); Bekoff and Sherman, "Reflections on Animal Selves."

16. Leslie Irvine, *If You Tame Me: Understanding Our Connection with Animals* (Philadelphia: Temple University Press, 2004), 240.

17. Karen McComb et al., "Matriarchs as Repositories of Social Knowledge in African Elephants," *Science* 292 (2001): 491–94.

18. Lucy A. Bates et al., "Elephants Classify Human Ethnic Groups by Odor and Garment Color," *Current Biology* 17 (2007): 1938–42.

19. Joyce H. Poole, Peter L. Tyack, Angela S. Stoeger-Horwath, and Stephanie Watwood, "Animal Behaviour: Elephants Are Capable of Vocal Learning," *Nature* 434 (2005): 455–56; Benjamin L. Hart, Lynette A. Hart, Michael McCoy, and C. R. Sarath, "Cognitive Behaviour in Asian Elephants: Use and Modification of Branches for Fly Switching," *Animal Behaviour* 62, no. 5 (2001): 839–47.

20. Arthur Glenberg, "Naturalizing Cognition: The Integration of Cognitive Science and Biology," *Current Biology* 16 (2005): R802–4.

21. Karen McComb, Lucy Baker, and Cynthia Moss, "African Elephants Show High Levels of Interest in the Skulls and Ivory of Their Own Species," *Biology Letters* 2, no. 1 (2006): 26–28.

22. Cynthia Moss, personal communication, January 14, 2009.

23. Benjamin L. Hart, Lynette Hart, and Noa Pinter-Wollman, "Large-Brains and Cognition: Where Do Elephants Fit In?" *Neuroscience and Biobehavioral Review* 32 (2008): 86–98.

24. Cynthia Moss and Martyn Colbeck, *Echo of the Elephants: The Story of an Elephant Family* (New York: William Morrow, 1992), 30.

25. Cynthia Moss, *Elephant Memories: Thirteen Years in the Life of an Elephant Family* (New York: William Morrow, 1992), 30.

26. Ibid., 601.

27. Kathleen Gobush, Ben Kerr, and Samuel Wasser, "Genetic Relatedness and Disrupted Social Structure in a Poached Population of African Elephants," *Molecular Ecology* 18 (2009): 722–34.

28. Pat Derby, personal communication, June 20, 2007.

29. Colbeck, in Moss and Colbeck, *Echo of the Elephants*, 9, 72.

30. "A Terror Monger Elephant Die in Ranchi," *Thaindian News*, January 27, 2008, http://www.thaindian.com/newsportal/india-news/a-terror-monger-elephant-die-in-ranchi_10014831.html.

31. George "Slim" Lewis and Byron Fish, *I Loved Rogues: The Life of an Elephant Tramp* (Seattle: Superior, 1978), 42 ff.

32. G. A. Bradshaw, Allan N. Schore, Janine L. Brown, Joyce Poole, and Cynthia Moss, "Elephant Breakdown," *Nature* 433 (2005): 807.

2. A Delicate Network

1. In Defense of Animals (IDA), "Dickerson Park Zoo," http://www.helpelephants.com/dickerson_zoo.html.

2. Michael J. Meaney, "Maternal Care, Gene Expression, and the Transmission of Individual Differences in Stress Reactivity Across Generations," *Annual Reviews of Neuroscience* 24 (2001): 1161–92; Robert M. Sapolsky, "Mothering Style and Methylation," *Nature Neuroscience* 7 (2004): 791–92; Ian C. G. Weaver et al., "Epigenetic Programming by Maternal Behaviour," *Nature Neuroscience* 7 (2004): 847–54.

3. Jack P. Shonkoff and Deborah Phillips, eds., *From Neurons to Neighborhoods: The Science of Early Childhood Development* (Washington, D.C.: National Academy Press, 2000), 388.

4. John Bowlby, *Attachment and Loss*, vol. 1, *Attachment* (New York: Basic, 1969).

5. John Bowlby, *A Secure Base: Parent-Child Attachment and Healthy Human Development* (New York: Basic, 1990), 65.

6. Judith R. Harris, "Social Behavior and Personality Development: The Role of Experiences with Siblings and with Peers," in *Origins of the Social Mind: Evolutionary Psychology and Child Development,* ed. Bruce J. Ellis and David F. Bjorklund (London: Guilford, 2005), 245–71.

7. Allan N. Schore, "Attachment, Affect Regulation, and the Developing Right Brain: Linking Developmental Neuroscience to Pediatrics," *Pediatrics in Review* 26 (2005): 204–17.

8. George Santayana, *The Life of Reason* (Amherst, N.Y.: Prometheus, 1998), 313.

9. Richard J. Davidson, "Affective Neuroscience and Psychophysiology: Toward a Synthesis," *Psychophysiology* 40 (2003): 655–65.

10. Bowlby, *Attachment and Loss,* vol. 1.

11. Heinz Düttman, Hans-Heiner Bergmann, Wiltraud Endländer, "Development and Behavior," in *Avian Growth and Development: Evolution Within the Altricial-Precocial Spectrum,* ed. J. Matthias Starck and Robert E. Ricklefs (New York: Oxford University Press, 1998), 223–45.

12. Phyllis C. Lee, unpublished data, February 2005; Keith Lindsay, personal communication, January 17, 2009.

13. Daphne Sheldrick, "Elephant Emotion," David Sheldrick Wildlife Trust Web site, http://www.sheldrickwildlifetrust.org/html/elephant_emotion.html.

14. As Dame Daphne writes (personal communication, October 31, 2008) about the need for rehabilitation centers: "At any age a young elephant duplicates its human counterpart in terms of age progression, so the orphans are not ready to leave their surrogate human family until about the age of 10 years or older. After all, one can't dump a human child of 10 in the middle of Russia and think that it is grown and can cope without adult supervision and guidance."

15. See the Trust's Web site, in particular http://www.sheldrickwildlifetrust.org/asp/orphans.asp and http://www.sheldrickwildlifetrust.org/about_us.asp.

16. Sheldrick, personal communication.

17. Several subspecies have also been identified, although some disagreement persists about species-subspecies division.

18. Myrtle Ryan and Peta Thornycroft, "Jumbos Mourn Black Rhino Killed by Poachers," *Independent Online,* November 18, 2007, http://www.iol.co.za/index.php?art_id=vn20071118084613595C384523.

19. Ritesh Joshi and Rambir Singh, "Unusual Behaviour of Asian Elephants in the Rajaji National Park, North-West India," *Gajah* 29 (2009): 32–34.

20. Baron Bror von Blixen-Finecke, *African Hunter* (New York: St. Martin's, 1986), 66.

21. Dalene Matthee, *Circles in the Forest* (London: Penguin, 1984), 23.

22. Cynthia J. Moss, *Elephant Memories: Thirteen Years in the Life of an Elephant Family* (New York: William Morrow 1988); Cynthia J. Moss, "The Demography of an

African Elephant (*Loxodonta africana*) Population in Amboseli, Kenya," *Journal of Zoology* 255 (2001): 145–56; Cynthia J. Moss and Joyce H. Poole, "Relationships and Social Structure of African Elephants," in *Primate Social Relationships: An Integrated Approach*, ed. Robert A. Hinde (Oxford: Blackwell Scientific Publications, 1983), 315–25.

23. Joyce H. Poole, "Announcing Intent: The Aggressive State of Musth in African Elephants," *Animal Behaviour* 37 (1989): 140–52.

24. Phyllis C. Lee, "Early Social Development in African Elephants," *National Geographic Research* 2 (1986): 394–401.

25. Phyllis C. Lee, "Early Social Development Among African Elephant Calves," *National Geographic Research* 2 (1986): 388–401; Phylis C. Lee, "Allomothering Among African Elephants," *Animal Behaviour* 35 (1987): 278–91.

26. Karen McComb et al., "Matriarchs as Repositories of Social Knowledge in African Elephants," *Science* 292 (2001): 491–94; Phyllis C. Lee, "Allomothering Among African Elephants," *Animal Behaviour* 35 (1987): 278–91; Phyllis C. Lee and Cynthia J. Moss, "Early Maternal Investment in Male and Female African Elephant Calves," *Behavioral Ecology and Sociobiology* 18 (1985): 353–61; Phyllis C. Lee and Cynthia J. Moss, "The Social Context for Learning and Behavioural Development Among Wild African Elephants," in *Mammalian Social Learning*, ed. Hilary O. Box and Kathleen R. Gibson (Cambridge: Cambridge University Press, 1999), 102–25; Moss and Poole, "Relationships and Social Structure."

27. McComb et al., "Matriarchs as Repositories of Social Knowledge."

28. Cynthia J. Moss and Harvey J. Croze, eds., *The Amboseli Elephants: A Long-Term Perspective on a Long-Lived Mammal* (Chicago: University of Chicago Press, in press).

29. From Dame Daphne (October 31, 2008): "What I have learnt in all the years that I have been hand-rearing wild species (and I have hand-reared most wild African mammals except a giraffe, and many birds as well, but never the primates) is that birds (and zebras) imprint on a human foster-parent, and therefore have difficulty in understanding their species or recognizing their peers. In the case of the zebra orphans, they imprint on the pattern of stripes of the mother, so whenever we get a zebra orphan, the Keeper wears a zebra striped jacked. Most other species (except the Primates) are possessed of a genetic memory which we call instinct, and exposure to a wild situation hones this genetic memory, or instinct, and they know exactly their place within the wild Kingdom. This explains the behavior of the crows mentioned in your article, and the tragic story of the Chimp, Billy, who was brought up as a human, and being a primate, had to learn everything in life, just as we do, his brain a blank until experience fills in the gaps."

30. G. A. Bradshaw, Theodora Capaldo, Gloria Grow, and Lorin Lindner, "Developmental Context Effects on Bicultural Post-Trauma Self Repair in Chimpanzees," *Developmental Psychology* (in press).

31. G. A. Bradshaw, Theodora Capaldo, Gloria Grow, and Lorin Lindner, "Building an Inner Sanctuary: Trauma-Induced Symptoms in Non-Human Great Apes," *Journal of Trauma and Dissociation* 9 (2008): 9–34.

32. Ibid.

33. "The Jessica Hippo Story," http://jessicahippo.co.za/index.php?option=com_content&task=view&id=1&Itemid=1.

34. Douglas Candland, *Feral Children and Clever Animals* (Oxford: Oxford University Press, 1995).

3. A Strange Kind of Animal

1. Lola González, Amended Confidential Forensic Assessment compiled by the defense counsel for Renaldo Devon McGirth, case number 2006-CF-002999-A-W. See the ongoing blog coverage of the incident and trial, Florida Issues, http://florida-issues.blogspot.com/2007_10_01_archive.html.

2. Rob Slotow and Gus van Dyk, "Role of Delinquent Young 'Orphan' Male Elephants in High Mortality of White Rhinoceroses in Pilanesberg, South Africa," *Koedoe* 44 (2001): 85–94.

3. Anna M. Whitehouse, "Managing Elephants: Lessons from Behavioural Studies," in *Elephant Conservation and Management in the Eastern Cape: Workshop Proceedings* (*TERU Report* 35), ed. G. Kerley, S. Wilson, and A. Massey (Port Elizabeth, South Africa: University of Port Elizabeth, 2002), 14–18.

4. Keith Lindsay, personal communication, January 16, 2009.

5. G. A. Bradshaw and Allan N. Schore, "How Elephants Are Opening Doors: Developmental Neuroethology, Attachment, and Social Context," *Ethology* 113 (2007): 426–36.

6. Bano Haralu, "Assam Elephants Face Shrinking Habitat," viewed June 29, 2005, at http://www.NDTV.com; Rob Slotow, personal communication, October 17, 2006.

7. Mark Owens and Delia Owens, *Cry of the Kalahari* (Boston: Houghton Mifflin, 1984).

8. B. C. Nigam, "Elephants of Jharkhand: Increasing Conflicts with Man," *Indian Forester* 128 (2002): 189–96.

9. "Elephants Kill 250 People in Eastern India," *Times of India*, June 5, 2008, http://timesofindia.indiatimes.com/Flora__Fauna/Elephants_kills_250_people_in_eastern_India/articleshow/3103373.cms.

10. "Human-Elephant Conflict Rampant in South Asia," OneWorld South Asia, November 21, 2008, http://southasia.oneworld.net/todaysheadlines/human-elephant-conflict-rampant-in-south-asia?searchterm=elephant; Haralu, "Assam Elephants Face Shrinking Habitat."

11. Debojeet Saikia, quoted in Haralu, "Assam Elephants Face Shrinking Habitat."

12. Meitamei Ole Dapash, "Coexisting in Kenya: The Human-Elephant Conflict," *Animal Welfare Institute* 521, no. 1 (2001), http://www.awionline.org/pubs/Quarterly/winter02/humanelephant.htm.

13. Martin Daly and Margo Wilson, "Crime and Conflict: Homicide in Evolutionary Psychological Perspective," *Crime and Justice* 22 (1997): 51–100.

14. Sean Nee, "The Great Chain of Being," *Nature* 435 (2005): 429.

15. C. G. Jung, *The Structure of the Psyche,* in *The Collected Works of C. G. Jung,* vol. 8, trans. Richard F. C. Hull (1931; Princeton: Princeton University Press, 1981), 152.

16. Robert A. Baron and Deborah R. Richardson, *Human Aggression,* 2nd ed. (New York: Plenum, 1994); James Gilligan, *Violence: Reflections on a National Epidemic* (New York: Vintage, 1996), 97.

17. C. G. Jung, *Conscious, Unconscious, and Individuation,* in Hull, *Collected Works,* vol. 9 (1939; Princeton: Princeton University Press, 1977), 280.

18. Jonathan Haidt, "The Moral Emotions," in *Handbook of Affective Sciences,* ed. Richard J. Davidson, Klaus R. Scherer, and H. Hill Goldsmith (Oxford: Oxford University Press, 2003), 856.

19. Daly and Wilson, "Crime and Conflict," 85.

20. Jane Goodall, *The Chimpanzees of Gombe: Patterns of Behavior* (Cambridge: Harvard University Press, 1986); Jane Goodall, *My Life with the Chimpanzees* (New York: Aladdin, 1996).

21. Antonio Preti, "Suicide Among Animals: A Review of Evidence," *Psychological Reports* 101 (2007): 831–48.

22. David Morris, "Animals and Humans, Thinking and Nature," *Phenomenology and the Cognitive Sciences* 4 (2005): 49–72.

23. Konrad Lorenz, *On Aggression* (New York: Harcourt, Brace, Jovanovich, 1963), 23, 25.

24. Cara Biega (producer), *Elephant Rage* (motion picture), National Geographic Corporation, 2005.

25. See Paola Cavalieri and Peter Singer, eds., *The Great Ape Project: Equality Beyond Humanity* (London: St. Martin's, 1993); Steven M. Wise, *Rattling the Cage: Toward Legal Rights for Animals* (Boston: Perseus, 2000).

26. "Fresh Steps to Curb Elephant Deaths," *Deccan Herald* (India), March 13, 2009, http://www.deccanherald.com/Content/Mar132009/scroll20090313123780.asp?section=updatenews.

27. Robert Hinde, *Animal Behaviour: A Synthesis of Ethology and Comparative Psychology* (New York: McGraw-Hill, 1966), 77.

28. Robert M. Sapolsky, *Why Zebras Don't Get Ulcers: An Updated Guide to Stress, Stress-Related Diseases, and Coping* (New York: Freeman, 1998), 419.

29. Viktor Frankl, *Man's Search for Meaning* (New York: Beacon, 1946), 135.

30. Michael Meaney, "Maternal Care, Gene Expression, and the Transmission of Individual Differences in Stress Reactivity Across Generations," *Annual Reviews of Neuroscience* 24 (2001): 1161–92.

31. Quoted in John Bowlby, *Attachment and Loss,* vol. 1, *Attachment* (New York: Basic, 1969), 331.

32. Meaney, "Maternal Care."

33. A. N. Schore, *Affect Regulation and Origin of the Self: The Neurobiology of Emotional Development* (Mahwah, N.J.: Erbaum, 1994); A. N. Schore, *Affect Dysregulation and Disorders of the Self* (Mahwah, N.J.: Erhbaum, 2003).

34. Lauren Slater, "Monkey Love," *Boston Globe,* March 21, 2004, www.boston.com/news/globe/ideas/articles/2004/03/21/monkey_love/.

35. See Deborah Blum, *The Monkey Wars* (New York: Oxford University Press, 1994). Also note that chimpanzee vivisection has been banned in most nations. The United States is one of the few remaining countries that has refused to do so.

36. Bowlby, *Attachment and Loss,* 1: xiii.

37. John Bowlby, *Separation: Anxiety and Anger* (New York: Basic, 1973), 371.

38. Judith Herman, *Trauma and Recovery: The Aftermath of Violence, from Domestic Abuse to Political Terror* (New York: Basic, 1997), 7.

39. See Sigmund Freud and J. Breuer, *On the Psychical Mechanism of Hysterical Phenomena: Preliminary Communication from Studies on Hysteria* (London: Penguin, 1893).

40. Herman, *Trauma and Recovery,* 20.

41. Siegfried Sassoon, Declaration Against the War, http://www.oucs.ox.ac.uk/ww1lit/education/ tutorials/intro/sassoon/declaration.html. Sassoon's complete text: "I am making this statement as an act of wilful defiance of military authority, because I believe that the War is being deliberately prolonged by those who have the power to end it. I am a soldier, convinced that I am acting on behalf of soldiers. I believe that this War, on which I entered as a war of defence and liberation, has now become a war of aggression and conquest. I believe that the purpose for which I and my fellow soldiers entered upon this war should have been so clearly stated as to have made it impossible to change them, and that, had this been done, the objects which actuated us would now be attainable by negotiation. I have seen and endured the sufferings of the troops, and I can no longer be a party to prolong these sufferings for ends which I believe to be evil and unjust. I am not protesting against the conduct of the war, but against the political errors and insincerities for which the fighting men are being sacrificed. On behalf of those who are suffering now I make this protest against the deception which is being practised on them; also I believe that I may help to destroy the callous complacency with which the majority of those at home regard the contriv-

ance of agonies which they do not, and which they have not sufficient imagination to realize."

42. Abram Kardiner, *The Traumatic Neuroses of War* (Washington, D.C.: National Research Council, 1941).

43. Herman, *Trauma and Recovery,* 33, 9.

44. J. C. Alexander and J. Colomy, "Traditions and Competition: Preface to a Post-Positivist Approach to Knowledge Cumulation," in *Metatheorizing,* ed. George Ritzer (Newberry Park, Calif.: Sage, 1992), 7–26. See also numerous books: Bessel A. van der Kolk, Alexander C. McFarlane, and Lars Weisaeth, eds., *Traumatic Stress: The Effects of Overwhelming Experience on Mind, Body, and Society* (New York: Guilford, 1996); John Bowlby, *Attachment and Loss,* vol. 1, *Attachment* (New York: Basic, 1969); Herman, *Trauma and Recovery;* Gilligan, *Violence;* Sapolsky, *Why Zebras Don't Get Ulcers;* John T. Cacioppo, Gary G. Berntson, Shelley E. Taylor, and Daniel L. Schacter, eds., *Foundations in Social Neuroscience* (Cambridge: MIT Press, 2002); and an almost endless list of journal articles and book chapters on diverse subjects: Rachel Yehuda et al., "Relationship Between Posttraumatic Stress Disorder Characteristics of Holocaust Survivors and Their Adult Offspring," *American Journal of Psychiatry* 155 (1998): 841–43; Bessel A. van der Kolk, "Posttraumatic Therapy in the Age of Neuroscience," in *Living with Terror, Working with Trauma: A Clinician's Handbook,* ed. Danielle Knafo (Lanham, Md.: Rowman and Littlefield 2004), 56–71; Bruce D. Perry et al., "Childhood Trauma, the Neurobiology of Adaptation, and 'Use-Dependent' Development of the Brain: How 'States' Become 'Traits,'" *Infant Mental Health Journal* 16 (2000): 271–91; Christoph P. Wiedenmayer, "Adaptations or Pathologies? Long-Term Changes in Brain and Behaviour after a Single Exposure to Severe Threat," *Neurosciences and Biobehavioural Reviews* 28 (2004): 1–12; Stephen J. Suomi, "How Gene-Environment Interactions Influence Emotional Development in Rhesus Monkeys," in *Nature and Nurture: The Complex Interplay of Genetic and Environmental Influences on Human Behaviour and Development,* ed. Cynthia García-Coll, Elaine L. Bearer, and Richard M. Lerner (Mahwah, N.J.: Lawrence Erlbaum, 2004), 35–51; Christina Heim and Charles B. Nemeroff, "The Impact of Early Adverse Experiences on Brain Systems Involved in the Pathophysiology of Anxiety and Affective Disorders," *Biological Psychiatry* 46 (1999): 1509–22; Henry Krystal, "Optimizing Affect Function in the Psychoanalytic Treatment of Trauma," in Knafo, *Living with Terror,* 283–96.

4. Deposited in the Bones

1. Judith Herman, *Trauma and Recovery: The Aftermath of Violence, from Domestic Abuse to Political Terror* (New York: Basic, 1997), 1.

2. "A History of the *Lunaes Montes*," Rwenzori.com, http://www.rwenzori.com/about_history.htm.

3. Evelyn L. Abe, quoted in G. A. Bradshaw, "Elephant Trauma and Recovery: From Human Violence to Trans-species Psychology," Ph.D. diss., Pacifica Graduate Institute, 2005, 214.

4. Abe, quoted ibid.

5. Evelyn L. Abe, "The Behavioural Ecology of Elephant Survivors in Queen Elizabeth National Park (QENP), Uganda," Ph.D. diss., University of Cambridge, 1994.

6. David Mabunda, Danie J. Pinaar, and Johan Verhoef, "The Kruger National Park: A Century of Management and Research," in *The Kruger Experience: Ecology and Management of Savanna Heterogeneity*, ed. Johan T. Du Toit, Kevin H. Rogers, and Harry C. Biggs (Washington, D.C.: Island, 2003), 3–21, quotations from 6.

7. Jane Carruthers, "Wilding the Farm or Farming the Wild? The Evolution of Scientific Game Ranching in South Africa from the 1960s to Present," *Transactions of the Royal Society of South Africa* 63 (2008): 160–81.

8. Magqubu Ntombela, "The Zulu Tradition," in *For the Conservation of the Earth*, ed. V. Martin (Boulder, Colo.: Fulcrum, 1988), 288–91, quoted in *Assessment of South African Elephant Management 2007*, http://www.elephantassessment.co.za/files/Assessment-of-South-African-Elephant-Management-2007.pdf, 397.

9. *Assessment of South African Elephant Management 2007*, 15.

10. Ian Leggett, *Uganda: OxFam Country Profile* (Oxford: Fountain, 2004), 32.

11. Quoted ibid., 88.

12. Ibid., 33.

13. "Uganda: Child 'Night Commuters,'" Amnesty International, http://www.amnesty.org/en/library/info/AFR59/013/2005.

14. The World Health Organization reports that efforts to curb the disease are starting to be successful in Uganda; "Uganda Reverses the Tide of HIV/AIDS," http://www.who.int/inf-new/aids2.htm.

15. David Blankenhorn, *Fatherless America: Confronting Our Most Urgent Social Problem* (New York: Harper Perennial, 1996), 45.

16. See Abe, "Behavioural Ecology of Elephant Survivors"; Joyce H. Poole, "The Effects of Poaching on the Age Structures and Social and Reproductive Patterns of Selected East African Elephant Populations," in *The Ivory Trade and the Future of the African Elephant*, vol. 2, Technical Reports, prepared by the Ivory Trade Review Group for the seventh CITES Conference of the Parties, 1989; Bill Jordan, *Who Cares for Planet Earth* (London: Alpha, 2001); Charles Siebert, "Are We Driving Elephants Crazy?" *New York Times Magazine*, October 6, 2006.

17. Delia Owens and Mark Owens, "Severe Poaching Alters Breeding Biology in African Savanna Elephants," *African Journal of Ecology* (2009, in press). First-birth

age has been correlated also with climate: "Females in Amboseli have a mean age of first birth that varies between thirteen and sixteen years, depending on the year it is calculated. In periods of good rainfall, the mean age of first birth drops, while in drought years it rises. The long-term average in Amboseli is just over fourteen." Keith Lindsay, personal communication, January 17, 2009.

18. C. A. Foley, S. Papageorge, and S. K. Wasser, "Noninvasive Stress and Reproductive Measures of Social and Ecological Pressures in Free-Ranging African Elephants," *Conservation Biology* 15 (2001): 1134–42.

19. "Uganda: Displaced First by War, Now by Elephants," *IRIN*, October 24, 2008, http://www.irinnews.org/Report.aspx?ReportId=81100.

20. Cuthbert Nzou, "Zimbabwe Govt Slaughters Elephants for Army," ZimOnline, January 9, 2009, http://www.zimonline.co.za/Article.aspx?ArticleId=4083; Joe Bavier, "Ivory Poachers Decimate Congo Elephant Population," Reuters News Service, August 22, 2008, http://www.reuters.com/article/environmentNews/idUSLM 40286220080822.

21. Patrick Omondi, Elphas Bitok, and Joachim Kagiri, "Managing Human-Elephant Conflicts: The Kenyan Experience," *Pachyderm* 36 (2004): 80–86.

22. Anna M. Whitehouse, "Tusklessness in the Elephant Population of the Addo Elephant National Park, South Africa," *Journal of Zoology* 257 (2002): 249–54; Silvester Nyakaana, Eve L. Abe, Peter Arctander, and Hans R. Siegismund, "DNA Evidence for Elephant Social Behaviour Breakdown in Queen Elizabeth National Park, Uganda," *Animal Conservation* 4 (2001): 231–37.

23. K. C. Panicker et al., "Captive Elephants and Human Conflicts in Kerala," in *Endangered Elephants: Past, Present, and Future*, ed. Jayantha Jayewardene (Colombo, Sri Lanka: Karunaratne, 2004), 21.

24. Bijoyananda Chowdhury, "Mela Shikar: A Dying Art," in Jayewardene, *Endangered Elephants*, 13–18.

25. Ibid.

26. Elke Riesterer, personal communication, March 11, 2005.

27. "Taming and Training of Elephants," *Sunday Observer* (Sri Lanka), http://www.sundayobserver.lk/2008/10/26/jun01.asp; Jacob Cheeran, personal communication, December 24, 2008.

28. Peter Jaeggi, quoted in Rhea Ghosh, *Gods in Chains* (Banglalore, India: Wildlife Rescue and Rehabilitation Center Press, 2005), 68.

29. Ibid.

30. Quoted ibid., 145.

31. Piyaporn Wongruang, "Elephants on the Loose in Bangkok," *Bangkok Post*, March 15, 2009, http://www.bangkokpost.com/news/investigation/13423/elephants-on-the-loose-in-bangkok.

32. Tunya Sukpanich, "Hunted in the Wild: Our National Symbol Is Losing the Freedom to Roam the Nation's Forests," *Bangkok Post,* March 8, 2009, http://www.bangkokpost.com/news/investigation/13000/hunted-in-the-wild.

33. Iain Douglas-Hamilton, "Keynote Address," in Jayewardene, *Endangered Elephants,* vi.

34. Cited in Ghosh, *Gods in Chains,* 159.

35. "Killer Elephant 'Osama' Shot Dead in Jharkhand (India)," Reuters News Service, May 31, 2008 http://in.reuters.com/article/topNews/idINIndia-33839920080531.

36. Apinya Wipatayotin, "Elephants' Future Truncated," *Bangkok Post,* March 13, 2009, http://www.bangkokpost.com/news/local/13242/elephants-future-truncated.

37. Martin Daly and Margo Wilson, "Crime and Conflict: Homicide in Evolutionary Psychological Perspective," *Crime and Justice* 22 (1997): 51–100.

38. Jeffrey C. Alexander, "Toward a Theory of Cultural Trauma," in *Cultural Trauma and Collective Identity,* ed. Jeffrey C. Alexander et al. (Berkeley: University of California Press, 2004), 7.

39. *Assessment of South African Elephant Management 2007,* 322.

40. "Madhbani Renews Park Bid," *New Vision,* April 12, 2006.

41. Jayantha Jayewardene, "Don't Make National Parks Ghettos for our Elephants," *Sunday Times,* August 5, 2007, http://www.sundaytimes.lk/070805/Plus/pls12.html.

42. "Human-Elephant Conflict Rampant in Southeast Asia," OneWorld South Asia, November 21, 2008, http://southasia.oneworld.net/todaysheadlines/human-elephant-conflict-rampant-in-south-asia?searchterm=elephant.

43. See Frantz Fanon, *The Wretched of the Earth,* trans. Constance Farrington (New York: Grove, 1965), 2.

5. Bad Boyz

1. Ros Clubb and Georgia Mason, *A Review of the Welfare of Elephants in European Zoos* (Horsham, U.K.: Royal Society for Prevention of Cruelty to Animals, 2002).

2. Chris T. Darimont et al., "Human Predators Outpace Other Agents of Trait Change in the Wild," *Proceedings of the National Academy of Science* 106 (2009), http://www.pnas.org/content/early/2009/01/12/0809235106.full.pdf+html.

3. Ibid.; Elizabeth A. Archie et al., "Behavioural Inbreeding Avoidance in Wild African Elephants," *Molecular Ecology* 16 (2007): 4138–48.

4. Mel Richardson, personal communication, March 8, 2009.

5. See John Allen, compiler, *The Essential Desmond Tutu* (Cape Town: David Phillip, 1997), 36.

6. "Enraged Bull Elephant Kills Herd's Matriarch in Safari Park," Associated Press, April 7, 2007, http://www.haaretz.com/hasen/spages/845495.html.

7. *Assessment of South African Elephant Management 2007,* http://www.elephantassessment.co.za/files/Assessment-of-South-African-Elephant-Management-2007.pdf; African Elephant Status Group, "African Elephant Status Report 2007," http://data.iucn.org/themes/ssc/sgs/afesg/aed/aesr2007.html.

8. South Africa Constitution, section 24:

Environment

Everyone has the right

a. to an environment that is not harmful to their health or well-being; and

b. to have the environment protected, for the benefit of present and future generations, through reasonable legislative and other measures that

   i. prevent pollution and ecological degradation;

   ii. promote conservation; and

   iii. secure ecologically sustainable development and use of natural resources while promoting justifiable economic and social development.

9. I. J. Whyte, R. J. van Aarde, and S. L. Pimm, "Kruger's Elephant Population: Its Size and Consequences for Ecosystem Heterogeneity," in *The Kruger Experience: Ecology and Management of Savanna Heterogeneity,* ed. Johan T. Du Toit, Kevin H. Rogers, and Harry C. Biggs (Washington, D.C.: Island, 2003), 332–48, statistics from 338.

10. Douglas H. Chadwick, *The Fate of the Elephant* (San Francisco: Sierra Club Books for Children, 1994), 431–32. Mike Cadman, a journalist from Zimbabwe, pointed out one technical inaccuracy in Read's account: "With a bolt-action rifle such as the .458 you have one round in the chamber and three in the magazine. The hunter has to use the bolt action of the rifle to move another round to the chamber after firing. In the case of the 7.62 mm military rifle—the Rhodesians used the Belgian-made FN and sometimes a South African copy called the R1—it has a twenty-round magazine and recocks automatically." Personal communication, November 4, 2008.

11. Chadwick, *Fate of the Elephant,* 432.

12. Rob Slotow and Gus van Dyk, "Role of Delinquent Young 'Orphan' Male Elephants in High Mortality of White Rhinoceroses in Pilanesberg, South Africa," *Koedoe* 44 (2001): 85–94; George Wittemyer et al., "Social Dominance, Seasonal Movements, and Spatial Segregation in the African Elephant: A Contribution to Conservation Behavior," *Behavioral Ecology and Sociobiology* 61 (2007): 1919–31.

13. Graeme Shannon et al., "African Elephant Home Range and Habitat Selection in Pongola Game Reserve, South Africa," *African Zoology* 41 (2006): 37–44; *Assessment of South African Elephant Management 2007,* 57.

14. Raman Sukumar, *The Living Elephant: Evolutionary Ecology, Behavior, and Conservation* (Oxford: Oxford University Press, 2003), 159.

15. "Botswana: On the Trunk Road," Telegraph.co.uk, March 30, 2001, http://www.telegraph.co.uk/travel/717661/On-the-trunk-road—Christopher-Munnion-tells-the-story-of-Randall-Moore-the-pioneer-of-elephant-back-safaris.html.

16. Slotow and van Dyk, "Role of Delinquent Young 'Orphan' Male Elephants."

17. G. A. Bradshaw and Allan N. Schore, "How Elephants Are Opening Doors: Developmental Neuroethology, Attachment, and Social Context," *Ethology* 113 (2007): 426–36.

18. Elsewhere, in areas such as Mikumi National Park, Tanzania, where the majority of resident elephants were lost to poaching before the ivory ban, Sam Wasser and Kathleen Gobush conducted a study on two hundred females and found that elephants in groups who lacked a matriarch or closely related relatives, particularly those living in areas where past poaching was most severe, had higher levels of stress hormones. These female elephants were statistically less apt to be pregnant or have an infant. Kathleen S. Gobush, B. M. Mutayoba, and Samuel K. Wasser, "Long-Term Impacts of Poaching on Relatedness, Stress Physiology, and Reproductive Output of Adult Female African Elephants," *Conservation Biology* 22 (2008): 1590–99.

19. D. C. Taylor, "Differential Rates of Cerebral Maturation Between Sexes and Between Hemispheres: Evidence from Epilepsy," *Lancet* 2 (1969): 140–42.

20. Bradshaw and Schore, "How Elephants Are Opening Doors."

21. Under natural populations, bull musth begins to occur between the ages of twenty-five and thirty years of age; its duration increases with age, lasting on average a few days in young males and up to four months for bulls older than forty; J. H. Poole, "Rutting Behaviour in African Elephants: The Phenomenon of Musth," *Behaviour* 102 (1987): 283–316; Heleen Druce et al., "The Effects of Mature Bull Introductions on Resident Bull's Group Size and Musth Periods: Phinda Private Game Reserve, South Africa," *South African Journal of Wildlife Research* 36 (2006): 133–37.

22. Bradshaw and Schore, "How Elephants Are Opening Doors."

23. Rob Slotow et al., "Older Bull Elephants Control Young Males," *Nature* 408 (2000): 425–26; Druce et al., "The Effects of Mature Bull Introductions."

24. American Psychiatric Association, *Diagnostic and Statistical Manual of Mental Disorders—Text Revision,* 4th ed. (Washington, D.C.: American Psychiatric Association, 1994), 467–69.

25. Quoted in Chadwick, *Fate of the Elephant,* 432.

26. Daphne Sheldrick, personal communication, November 2, 2008.

27. However, AERP researcher Cynthia Moss has tracked details of individuals from birth through adulthood and could speak to their differences and patterns relative to life changes that may occur.

28. Daphne Sheldrick, personal communication, October 31, 2008.

29. Ibid.

30. Sentencing Order in the Circuit Court of the Fifth Judicial Circuit, in and for Marion County, Florida, case no. 06–2999-CF-A-W, *State of Florida vs. Renaldo Devon McGirth;* see also Mabel Pérez, "Two Will Go to Trial in the Villages Slaying Case: Defense Denied Request for More Time to Investigate Role of Victim's Daughter," *Star Banner,* October 26, 2007, http://www.ocala.com/article/20071026/NEWS/210260350/1001/NEWS01&MaxW=270&MaxH=200.

31. Lola González, personal communication, October 28, 2008.

32. Lola González, Amended Forensic Assessment, interview dates October 21, 2007–February 11, 2008.

33. González, personal communication.

34. González asserts that this scenario is common with the children with whom she works. Renaldo's mother had told him "countless times" that one of her boyfriends was his father. She would tell him that his father had promised to pick up Renaldo for an outing together or to bring him a birthday present, but these promises were never fulfilled. González says that Renaldo describes these incidents, the hurt and feeling of abandonment, as if they had happened just the day before. Ibid.

35. Renaldo McGirth mitigation documents.

36. González, personal communication.

37. "Life in the Balance Conference," Pro Bono.net, http://www.probono.net/deathpenalty/calendar/event.227992-Life_in_the_Balance_Conference.

38. David Blankenhorn, *Fatherless America: Confronting Our Most Urgent Social Problem* (New York: Harper Perennial, 1996), 1.

39. Aaron Kipnis, *Angry Young Men: How Parents, Teachers, Counselors, Can Help "Bad Boys" Become Good Men* (San Francisco: Jossey-Bass, 1999), 37.

40. Robert Bly, *Iron John: A Book About Men* (New York: Vintage, 1992).

41. Steve Biddulph, *Raising Boys* (Berkeley: Celestial Arts, 1997), 100.

42. Michael Flood, "Fatherhood and Fatherlessness," discussion paper no. 59 (Manuka: Australia Institute, 2003), 25; Raymond A. Knight and Robert A. Prentky, "The Developmental Antecedents and Adult Adaptations of Rapist Subtypes," *Criminal Justice and Behavior* 14 (1987), 403–26.

43. Quoted in Kipnis, *Angry Young Men,* 152.

44. James Gilligan, *Violence: Reflections on a National Epidemic* (New York: Vintage, 1996), 45.

45. See Vincent J. Felitti et al., "Relationship of Childhood Abuse and Household Dysfunction to Many of the Leading Causes of Death in Adults: The Adverse Childhood Experience (ACE) Study," *American Journal of Preventive Medicine* 14 (1998):

254–58; Felicity deZulueta, "Inducing Traumatic Attachment in Adults with a History of Child Abuse: Forensic Applications," *British Journal of Forensic Practice* 8, no. 3 (2006): 4–15.

46. Mark Hunter, "Fathers Without *Amandla:* Zulu-speaking Men and Fatherhood," in *Baba: Men and Fatherhood in South Africa,* ed. Linda Richter and Robert Morrell (Pretoria: Human Sciences Research Council, 2006), 99–107.

47. Ibid., 101.

48. Mamphela Ramphele and Linda Richter, "Migrancy, Family Dissolution, and Fatherhood," in Richter and Morrell, *Baba,* 99–107.

49. Stacey L. Williams et al., "Multiple Traumatic Events and Psychological Distress: The South Africa Stress and Health Study," *Journal of Traumatic Stress* 20 (2007): 845–55.

50. Kipnis, *Angry Young Men,* 152; Nancy Rugart (director), *Oh Father . . . Where Art Thou?* (motion picture), 2007.

6. Elephant on the Couch

1. Ros Clubb and Georgia J. Mason, *A Review of the Welfare of Elephants in European Zoos* (Horsham, U.K.: Royal Society for the Prevention of Cruelty to Animals, 2002); Ros Clubb et al., "Compromised Survivorship in Zoo Elephants," *Science* 322 (2008): 1649; see also Debra L. Forthman et al., eds. *An Elephant in the Room: The Science and Well-Being of Elephants in Captivity* (North Grafton, Mass.: Tufts Center for Animals and Public Policy, 2009).

2. G. A. Bradshaw, "Inside Looking Out: Neuroethological Compromise Effects on Elephants in Captivity," in Forthman et al., *Elephant in the Room,* 55–68.

3. In the period between the Civil War and World War I, all male elephants in U.S. circuses were killed. See Shana Alexander, *The Astonishing Elephant* (New York: Random House, 2000), chapter 7, "The Disappearance."

4. Humane Scorecard: A Project of the Humane Society and Funds for Animals, 2003, www.fund.org/pdfs/scorecard_108th_midterm.pdf, 2; G. A. Bradshaw, "Elephants in Captivity: Analysis of Practice, Policy, and the Future," *Society and Animals* 12 (2007): 143–52.

5. Naralie Neysa Alund, "Lawsuit Alleges Elephant Abuse," Bradenton Herald .com, October 21, 2008, http://www.elephants.com/media/Bradenton Herald_10_21_08.htm.

6. Lisa Kane, Debra Forthman, and David Hancocks, eds., *Best Practices by the Coalition for Captive Elephant Wellbeing,* 2005, www.elephantcare.org/protodoc_files/new%2006/CCEWBCoreBestPractices.2.pdf, 21–25.

7. Oregon Zoo Medical Records retrieved from In Defense of Animals (IDA), November 2, 2008, http://www.helpelephants.com/oregon_health.html.

8. Alan Roocroft and D. A. Zoll, *Managing Elephants* (Ramona, Calif.: Fever Tree, 1994), 104.

9. Quoted in In Defense of Animals, *A Report on the Failures in Care and Elephant Management at the Six Flags Amusement Park* (San Rafael, Calif.: In Defense of Animals, 2007), www.savezooelephants.org/records/six_flags/report_070628.pdf.

10. Michael Schmidt, *Jumbo Ghosts: The Dangerous Life of Elephants in the Zoo* (Philadelphia: Xlibris, 2002), 19; Ros Clubb and Georgia J. Mason, *A Review of the Welfare of Elephants in European Zoos* (Horsham, U.K.: Royal Society for the Prevention of Cruelty to Animals, 2002): Janine L. Brown, N. Wielebnowski, and Jacob V. Cheeran, "Evaluating Stress and Well-Being in Elephants: Are We Doing All We Can to Make Sure Elephants Are Well Cared For?" *Science India* 7 (2000): 88–93.

11. Elephants used in the logging industry are under duress and often sustain injury from the dangerous work, but when they are not working they are permitted to roam and to live in the forest, socialize, and mate, even have families under more "natural" conditions. Researchers believe that the greater space and socialization decreases the stress sufficiently to affect overall health positively relative to zoos. "Elephants 'Die earlier in Zoos,'" BBC News, December 8, 2008, http://news.bbc.co.uk/2/hi/science/nature/7777413.stm.

12. Clubb et al., "Compromised Survivorship."

13. Mel Richardson, testimony for In Defense of Animals (IDA), http://www.helpelephants.com/expert.html.

14. Quoted in Help Elephants in Zoos, http://www.helpelephants.com/oregon_zoo.html.

15. Elephant Birth Protocol, Riddle's Elephant and Wildlife Sanctuary, in Louisville Zoological Garden, *Elephant Husbandry Manual,* updated February 2007, 79.

16. Clubb et al., "Compromised Survivorship."

17. James Gilligan, *Violence: Reflections on a National Epidemic* (New York: Vintage, 1996), 37.

18. *Pica* derives from the Latin name for the magpie, *Pica pica,* because the bird is thought to have a constant desire to peck and place objects in its mouth.

19. Chris Cantor, *Evoution and Posttraumatic Stress: Disorders of Vigilance and Defence* (Sussex, U.K.: Routledge, 2005).

20. "Arguments Begin on Elephant Care," *Washington Post,* February 5, 2009, http://www.washingtonpost.com/wp-dyn/content/article/2009/02/04/AR2009020403382.html?hpid=moreheadlines.

21. Joanna Cattanach, "Dallas Zoo to Add 10-acre African Savanna Attraction with Lions, Elephants Next Spring," *Dallas Morning News,* March 23, 2009, http://www.dallasnews.com/sharedcontent/dws/news/localnews/stories/032309dnmetzooexpand.3766242.html?ocp=2.

7. The Sorrow of the Cooking Pot

1. Desmond Tutu, *No Future Without Forgiveness* (New York: Image Doubleday, 1997), 35.

2. Frantz Fanon, *Black Skin, White Masks,* trans. Richard Philcox (New York: Grove, 2008), 69.

3. Robert Jay Lifton, "Is Hiroshima Our Text?" in *Living with Terror, Working with Trauma: A Clinician's Handbook,* ed. Danielle Knafo (Lanham, Md.: Bowman and Littlefield, 2005), 111.

4. Elie Wiesel, *Night* (New York: Hill and Wang, 2006), 109.

5. Tutu, *No Future Without Forgiveness,* 249.

6. Peter Homans, *Symbolic Loss: The Ambiguity of Mourning and Memory at Century's End* (Charlottesville: University of Virginia Press), 353.

7. Daphne Sheldrick, personal communication, June 10, 2007.

8. Edward Tick, *War and the Soul: Healing Our Nation's Veterans from Post-Traumatic Stress Disorder* (Wheaton, Ill.: Quest, 2005), 16.

9. Quoted in Diego Gambetta, "Primo Levi's Last Moments," *Boston Review,* April–May 2004, http://bostonreview.net/BR24.3/gambetta.html.

10. Dianna Ortiz, *Blindfold's Eyes: My Journey from Torture to Truth* (New York: Orbis), 476.

11. Judith Herman, *Trauma and Recovery: The Aftermath of Violence, from Domestic Abuse to Political Terror* (New York: Basic Books, 1997), 32.

12. Quoted in Paul Hickman, "But You Weren't There," in *Post-Traumatic Stress Disorders: A Handbook for Clinicians,* ed. Tom Williams (Cincinnati: Disabled American Veterans, 1987), 193–207.

13. Zoos have also been criticized for their restrictive use of the medical model in care of elephants and other animals; see Lisa Kane, "Contemporary Zoo Elephant Management: Captive to a 19th-Century Vision," in *An Elephant in the Room: The Science and Well-Being of Elephants in Captivity,* ed. Debra L. Forthman et al. (North Grafton, Mass.: Tufts Center for Animals and Public Policy, 2009), 87–98.

14. Tick, *War and the Soul,* 169.

15. In their study of postwar Europe, the British and German scholars Geoffrey Gorer and Alexander and Marguerite Mitscherlichs, respectively, as well as the Amer-

ican scholar Ernest Becker, have written extensively about the psychological denial and its effects in history.

16. Wiesel, *Night,* x, ix.

17. Patrick J. Bracken, Joan E. Giller, and Derek Summerfield, "Rethinking Mental Health Work with Survivors of Wartime Violence and Refugees," *Journal of Refugee Studies* 10 (1997): 431–42; Patrick J. Bracken, Joan E. Giller, and Derek Summerfield, "Psychological Responses to War and Atrocity: The Limitations of Current Concepts," *Social Science and Medicine* 40 (1995): 1073–82; Patrick Bracken, Joan Giller, and Derek Summerfield, "War and Mental Health: A Brief Review," *British Medical Journal* 321 (2000): 232–35.

18. Laurence Kirmayer, "Psychotherapy and the Cultural Concept of the Person," *Transcultural Psychiatry* 44 (2007): 232–57.

19. Homans, *Symbolic Loss: The Ambiguity of Mourning and Memory at Century's End,* 17.

20. Tick, *War and the Soul,* 169.

21. Hannah Arendt, *Eichmann in Jerusalem: A Report on the Banality of Evil* (New York: Penguin Classics, 1992).

22. Tick, *War and the Soul,* 169.

23. Ibid.

24. Herman, *Trauma and Recovery,* 7.

25. Ignacio Martín-Baró, *Writings for a Liberation Psychology,* ed. A. Aron and S. Corne (Cambridge: Harvard University Press, 1994), 25.

26. Deogratias Bagilishya, "Mourning and Recovery from Trauma: In Rwanda, Tears Flow Within," *Transcultural Psychiatry* 37 (2000): 337–53.

27. Ibid., 340.

28. G. A. Bradshaw, Theodora Capaldo, Lorin Lindner, and Gloria Grow, "Developmental Context Effects on Bicultural Post-Trauma Self Repair in Chimpanzees," *Developmental Psychology,* in press; G. A. Bradshaw, Theodora Capaldo, Lorin Lindner, and Gloria Grow, "Building an Inner Sanctuary: Trauma-Induced Symptoms in Non-Human Great Apes," *Journal of Trauma and Dissociation* 9 (2008): 9–34.

8. The Biology of Forgiveness

1. Chen Collins et al., "Health Status and Mortality in Holocaust Survivors Living in Jerusalem 40–50 Years Later," *Journal of Traumatic Stress* 17 (2004): 403.

2. List compiled from public records by In Defense of Animals, courtesy Catherine Doyle.

3. Michael E. McCullough, "Forgiveness: Who Does It and How Do They Do It?" *Current Directions in Psychological Science* 10 (2001): 194–97.

4. Deogratias Bagilishya, "Mourning and Recovery from Trauma: In Rwanda, Tears Flow Within," *Transcultural Psychiatry* 37 (2000): 337–53.

5. Nancy Scheper-Hughes, "Undoing: Social Suffering and the Politics of Remorse in the New South Africa," *Social Justice* 25 (1998): 114–42; Nancy Scheper-Hughes and Philippe I. Bourgois, eds., *Violence in War and Peace: An Anthology* (Malden, Mass.: Wiley-Blackwell), 2003.

6. Gordon S. Gallup, "Self-Recognition in Primates: A Comparative Approach in Bidirectional Properties of Consciousness," *American Psychologist* 32 (1977): 329–38; John Bowlby, *Attachment and Loss,* vol. 1, *Attachment* (New York: Basic, 1969).

7. Tom F. D. Farrow et al., "Quantifiable Change in Functional Brain Response to Empathic and Forgivability Judgments with Resolution of Posttraumatic Stress Disorder," *Psychiatry Research: Neuroimaging* 140 (2005): 45–53.

8. Allan N. Schore, *Affect Dysregulation and Disorders of the Self* (Mahwah, N.J.: Erhbaum, 2003).

9. Felicity de Zulueta, "Inducing Traumatic Attachment in Adults with a History of Child Abuse: Forensic Applications," *British Journal of Forensic Practice* 8, no. 3 (2006): 4–15.

10. Henry Krystal, "Optimizing Affect Function in the Psychoanalytic Treatment of Trauma," in *Living with Terror, Working with Trauma: A Clinician's Handbook,* ed. Danielle Knafo (Lanham, Md.: Bowman and Littlefield, 2004), 283–96, quotation from 286.

11. John Bowlby, *A Secure Base: Parent-Child Attachment and Healthy Human Development* (New York: Basic, 1990), 64.

12. Patricia Greenfield et al., "Cultural Pathways Through Universal Development," *Annual Reviews of Psychology* 54 (2003): 461–90. See also, for example, Laurence Kirmayer, "Psychotherapy and the Cultural Concept of the Person," *Transcultural Psychiatry* 44 (2007): 232–57.

13. Laurence Kirmayer, "Beyond the 'New Cross-Cultural Psychiatry': Cultural Biology, Discursive Psychology, and the Ironies of Globalization," *Transcultural Psychiatry* 43 (2006): 126–44.

14. Soo See Yeo, "Bonding and Attachment of Australian Aboriginal Children," *Child Abuse Review* 12 (2004): 292–304.

15. Allan N. Schore, "Attachment, Affect Regulation, and the Developing Right Brain: Linking Developmental Neuroscience to Pediatrics," *Pediatrics in Review* 26 (2005): 204–17.

16. Andre B. Newberg et al., "The Nueropsychological Correlates of Forgiveness," in *Forgiveness: Theory, Research, and Practice,* ed. Michael E. McCullough, Kenneth I. Pargament, and Carl E. Thoresen (New York: Guilford, 2001), 99–110.

17. V. Igreja, "'*Why Are There So Many Drums Playing Until Dawn?*' Exploring the

Role of *Gamba* Spirits and Healers in the Post-War Recovery Period in Gorongosa, Central Mozambique," *Transcultural Psychiatry* 40 (2003): 459–87.

18. Joachim Kadima Kadiangandu, Mélanie Gauché, Geneviève Vinsonneau, and Etienne Mullet, "Conceptualizations of Forgiveness: Collectivist-Congolese Versus Individualist-French Viewpoints," *Journal of Cross-Cultural Psychology* 38 (2007): 432–37.

19. "Ndume Visits Again After a Long Absence," Malaika and Ndume Diaries, David Sheldrick Wildlife Trust, 7/15/2007, http://www.sheldrickwildlifetrust.org/updates/updates.asp?ID=124.

20. Daphne Sheldrick, personal communication, February 2008.

21. Ibid.

22. Ibid.

23. Sheldrick, personal communication, February 2008.

24. Krystal, "Optimizing Affect Function," 67.

25. Bowlby, *Attachment*, 223.

26. Shai Lavi, "The Jews Are Coming": Vengence and Revenge in Post-Nazi Europe," *Law, Culture, and Humanities* 1 (2005): 282–301.

27. John Bishop and Robert C. Lane, "The Dynamics and Dangers of Entitlement," *Psychoanalytic Psychology* 19 (2002): 739–58.

28. Newberg et al., "The Nueropsychological Correlates of Forgiveness."

9. Am I an Elephant?

1. Carol Buckley, personal communication, April 6, 2006.

2. Judith Herman, *Trauma and Recovery: The Aftermath of Violence, from Domestic Abuse to Political Terror* (New York: Basic, 1997), 119.

3. John Briere and Joseph Spinazzola, "Phenomenology and Psychological Assessment of Complex Posttraumatic States," *Journal of Traumatic Stress* 18 (2005): 401–12.

4. Herman, *Trauma and Recovery*, 115; G. A. Bradshaw, "Inside Looking Out: Neurobiological Compromise Effects in Elephants in Captivity," in *An Elephant in the Room: The Science and Well-Being of Elephants in Captivity*, ed. Debra L. Forthman et al. (North Grafton, Mass.: Tufts Center for Animals and Public Policy, 2009), 55–68.

5. Michael Schmidt, *Jumbo Ghosts: The Dangerous Life of Elephants in the Zoo* (Philadelphia: Xlibris, 2002), 114.

6. "Tarra's Story," Elephant Sanctuary, http://www.elephants.com/tarra/tarra.htm.

7. Amy Mayers, personal communication, August 4, 2008.

8. "Ned's Bio," Elephant Sanctuary, http://www.elephants.com/Ned/ned_bio.htm.

9. "Once Upon a Time There Was a Baby Elephant . . . ," Elephant Sanctuary, http://www.elephants.com/winkie/winkiebio.htm.

10. "Is It Really You? Shirley and Jenny Together Again!" Elephant Sanctuary, http://www.elephants.com/shirley/shirleypic3.htm; "Elephant Biographies: Jenny's," www.elephants.com/jenny/jenny.htm.

11. Herman, *Trauma and Recovery*, 91–92.

12. Elie Wiesel, *Night* (New York: Hill and Wang, 2006), 36.

13. "Shirley's Independence Day," Elephant Sanctuary Web site, http://www.elephants.com/shirley/shirleyarrives.htm.

14. Bessel A. van der Kolk, Alexander C. McFarlane, and Lars Weisaeth, eds., *Traumatic Stress: The Effects of Overwhelming Experience on Mind, Body, and Society* (New York: Guilford, 1996), 214.

15. The Elephant Sanctuary in Tennessee Stands Ready to Rescue 6 Ailing Circus Elephants Including . . . ," AllBusiness, June 24, 2004, http://www.allbusiness.com/environment-natural-resources/ecology-environmental/5663235-1.html; "The Hawthorn Elephant Rescue," Trunklines, www.elephants.com/pdf/Trunklines_Spring06.pdf.

16. Buckley, personal communication.

17. Ibid.

18. Viktor Frankl, *Man's Search For Meaning* (New York: Simon and Schuster, 1985), 135.

19. Herman, *Trauma and Recovery*, 83.

20. Ibid.

21. Nibha Namboodiri, compiler and ed., K. C. Panicker and Jacob V. Cheeran, technical eds., *Practical Elephant Management: A Handbook for Mahouts*, section VI (Coimbatore, India: Elephant Welfare Association, 1997), http://www.elephantcare.org/mantitle.htm. Both practices evoke images of the Raj, under which the Indian Army, in service to the British, was sent to fight other Indians. "Kunki Elephants to Guide Jumbos Disturbing Nadora," Indianews, July 13, 2005, www.elephant-news.com/index.php?id=545.

22. Herman, *Trauma and Recovery*, 94.

23. Quoted in Buckley, personal communication.

24. Ibid., 119.

25. William G. Niederland, "Clinical Observations on the Survivor Syndrome," *International Journal of Psychoanalysis* 49 (1968): 313–35, quoted ibid., 114.

26. Ibid., 93–94.

27. Edward Tick, *War and the Soul: Healing Our Nation's Veterans from Post-Traumatic Stress Disorder* (Wheaton, Ill.: Quest, 2005), 16.

28. "Queenie, 1959–2008," Elephant Sanctuary, http://www.elephants.com/queenie/queenie_inmemory.htm.

29. Herman, *Trauma and Recovery*, 156.

30. David Spurr, *The Rhetoric of the Empire: Colonial Discourse in Journalism, Travel Writing, and Imperial Administration* (Durham: Duke University Press, 1993), 15.

31. Herman, *Trauma and Recovery*, 162.

32. Joyce Poole, keynote address at the 22nd Annual Conference of Elephant Managers Association, Orlando, Florida, 2001.

33. Buckley, personal communication.

34. "Billie," Elephant Sanctuary, http://www.elephants.com/billie/billie_bio.htm.

35. Buckley, personal communication.

36. Herman, *Trauma and Recovery*, 178.

37. "Flora's Bio," Elephant Sanctuary, http://www.elephants.com/flora/flora_bio.htm. Until January 11, 2009, three African elephants lived at the Sanctuary. However, Zula unexpectedly died, leaving Flora and Tange. "Zula, 1975–January 11, 2009," Elephant Sanctuary, http://www.elephants.com/zula/zulastart_inmemory.htm.

38. Buckley, personal communication.

39. Bradshaw, "Inside Looking Out"; Herman, *Trauma and Recovery*.

40. Quoted in Rhea Ghosh, *Gods in Chains* (Banglalore, India: Wildlife Rescue and Rehabilitation Center Press, 2005), 112.

41. G. A. Bradshaw, "Elephants in Circuses: Analysis of Practice, Policy, and the Future," policy paper (Ann Arbor, Mich.: Animals and Society Institute, 2007), 1–48.

42. John Bowlby, *Attachment and Loss*, vol. 3, *Loss: Sadness and Depression* (London: Hogarth, 1980), 71.

43. See Pat Ogden, Kekuni Minton, and Clare Pain, *Trauma and the Body: A Sensorimotor Approach to Psychotherapy* (New York: Norton, 2006).

44. Buckley, personal communication.

45. Stephen Porges, "The Polyvagal Theory: Phylogenetic Contributions to Social Behavior," *Physiology and Behaviour* 79 (2003): 503–13; Ogden, Minton, and Pain, *Trauma and the Body*.

46. Ogden, Minton, and Pain, *Trauma and the Body*.

47. Sylvia Plath, *The Bell Jar* (New York: Bantam, 1972), 207.

48. Sandra de Rek, personal communication, May 6, 2006.

49. "Elephant Kills Handler in Tennessee Sanctuary," *USA Today*, July 21, 2006, http://www.usatoday.com/news/nation/2006–07–21-elephant-attack_x.htm; G. A. Bradshaw and Lorin Linder, "Posttraumatic Stress and Elephants in Captivity," http://www.elephants.com/joanna/Bradshaw&Lindner_PTSD-rev.pdf.

50. Cathy Caruth, ed., *Trauma: Explorations in Memory* (Baltimore: Johns Hopkins University Press, 1995), 3.

51. "Delhi," Elephant Sanctuary, http://www.elephants.com/delhi/delhi_bio.htm.

52. "Delhi," Elephant Sanctuary, http://www.elephants.com/delhi/delhi_tributes.php.

10. Speaking in Tongues

1. Listening Hands, www.listening-hands.org.

2. Aldabra Tortoise, Oakland Zoo, http://www.oaklandzoo.org/animals/reptiles/aldabra-tortoise/.

3. Linda Tellington-Jones and Sybil Taylor, *The Tellington TTouch: A Revolutionary Natural Method to Train and Care for Your Favorite Animal* (New York: Penguin, 1995).

4. Elke Riesterer, personal communication, March 22, 2007.

5. Ray Ryan, personal communication, December 15, 2008.

6. Elie Wiesel, *Night* (New York: Hill and Wang, 2006), 85.

7. Carol Buckley, personal communication, April 6, 2006.

8. Janine Brown, "Reproductive Endocrine Monitoring of Elephants: An Essential Tool for Assisting Captive Management," *Zoo Biology* 19 (2000): 347–67; Thomas B. Hildebrandt et al., "Ultrasonography of the Urogenital Tract in Elephants (*Loxodonta africana* and *Elephas maximus*): An Important Tool for Assessing Male Reproductive Function," *Zoo Biology* 19 (2000): 333–45.

9. See Hildebrandt et al., "Ultrasonography"; "The Elephant's Guide to Sex," BBC, April 19, 2009, http://www.bbc.co.uk/sn/tvradio/programmes/horizon/broadband/tx/elephant/highlights.

10. Elke Riesterer, personal communication, June 14, 2007.

11. The former South African prisoner wishes to remain anonymous.

12. Judith Herman, *Trauma and Recovery: The Aftermath of Violence, from Domestic Abuse to Political Terror* (New York: Basic, 1997).

13. Elke Riesterer, personal communication, June 17, 2007.

14. Celeste Wiser, personal communication, August 23, 2007.

15. Dori Laub, "Truth and Testimony: The Process and the Struggle," in *Trauma: Explorations in Memory*, ed. C. Caruth (Baltimore: Johns Hopkins University Press, 1995), 61–75.

16. C. G. Jung, *The Practice of Psychotherapy*, in *The Collected Works of C. G. Jung*, vol. 16, trans. R. F. C. Hull, ed. Herbert Edward Read, Michael Fordham, Gerhard Adler (Princeton: Princeton University Press, 1973), 132.

17. Sigmund Freud, "Project for a Scientific Psychology," *The Standard Edition of*

the Complete Psychological Works of Sigmund Freud, vol. 1, ed. James Strachey, (1895; London: Hogarth, 1950), 295–397.

18. Riesterer, personal communication, June 17, 2007.

19. Allan N. Schore, "Attachment, Affect Regulation, and the Developing Right Brain: Linking Developmental Neuroscience to Pediatrics," *Pediatrics in Review* 26, no. 6 (2005): 204–17.

20. Michael Mathew, "The Body as Instrument," *Journal of the British Association of Psychotherapists* 35 (1998): 17–36.

21. Allan N. Schore, "A Neuropsychoanalytic Commentary on 'Body Rhythms and the Unconscious: Toward an Expanding of Clinical Attention,'" Psychopanalytic Dialogues, Psy Broadcasting Company, 2005, 23, http://psybc.com/pdfs/library/Neuropsa_BodyRythymsUcs_Schore.pdf.

22. Ryan, personal communication.

23. Riesterer, personal communication, June 14, 2007.

24. Pat Ogden, Kekuni Minton, and Clare Pain, *Trauma and the Body: A Sensorimotor Approach to Psychotherapy* (New York: Norton, 2006).

25. Riesterer, personal communication, June 14, 2007.

11. Where Does the Soul Go?

1. Jane Fritsch, "Elephants in Captivity: A Dark Side," *Los Angeles Times*, October 15, 1988.

2. "No Charge Filed in Elephant Discipline Case," news release, Office of the City Attorney of San Diego John W. Witt, June 30, 1988.

3. Fred Kurt and Marion Garai, *The Asian Elephant in Captivity: A Field Study* (New Delhi: Foundation, Cambridge University Press, 2007).

4. San Diego County Humane Society & S.P.C.A., State Humane Officers Investigation Reports Case 88-1853, Violation 597(b)PC.

5. Interim Hearing on San Diego Zoological Society: The Care and Handling of Animals and Other Management Issues, July 29, 1988, California Legislature, Senate Committee on Natural Resources and Wildlife, Senator Dan McCorquodale, Chairman, 48–49.

6. George "Slim" Lewis and Byron Fish, *I Loved Rogues: The Life of an Elephant Tramp* (Seattle: Superior, 1978), 77, 28–30.

7. Lisa Kane, Debra Forthman, and David Hancocks, "Optimal Conditions for Captive Elephants: A Report by the Coalition for Captive Elephant Wellbeing," Coalition for Captive Elephant Well-Being, 2005, http://www.elephantcare.org/protodoc_files/new%2006/CCEWBOptimalConditionspdf.2.pdf, 3y.

8. Interim Hearing, 53.

9. San Diego County Humane Society & S.P.C.A., State Humane Officers Investigation Reports.

10. Ibid.

11. Ibid.

12. Ray Ryan, personal communication, December 2, 2008.

13. Ibid.

14. Nedra Pickler, "Circus CEO Says Elephants Are Struck, But Not Hurt," Associated Press, March 3, 2009, http://www.google.com/hostednews/ap/article/ALeqM5iIvnEu-D_CqeGOboBYNAvQvp8EvQD96MST600.

15. Eric Scigliano, *Love, War, and Circuses: The Age-Old Relationship Between Elephants and Humans* (New York: Houghton Mifflin, 2002), 277.

16. Lewis and Fish, *I Loved Rogues*, 96.

17. Scigliano, *Love, War, and Circuses*, 262.

18. "Zoo's Elephant Hansa Dies Suddenly," Woodland Park Zoo press release, June 8, 2007, http://www.zoo.org/pressroom/pr/2007/pr06_08_2007.htm.

19. Leon Stafford, "Zoo Atlanta's Pregnant Elephant Dies," *Atlanta Journal-Constitution*, October 28, 2008, http://www.ajc.com/living/content/metro/atlanta/stories/2008/10/28/atlanta_elephant_dies.html?cxntlid=inform_artr; "Precocious Elephant Dies, Zoo Mourns," CBS News, November 15, 2008, http://www.cbsnews.com/stories/2008/11/15/national/main4607126.shtml.

20. Quoted in Scigliano, *Love, War, and Circuses*, 262.

21. Ray Ryan, personal communication, December 2, 2008.

22. See Randy Malamud, *Poetic Animals and Animal Souls* (New York: Palgrave McMillan, 2003); Scigliano, *Love, War, and Circuses;* Janet M. Davis, *The Circus Age: Culture and Society Under the American Big Top* (Chapel Hill: University of North Carolina Press, 2002).

23. Susan Griffin, *A Chorus of Stones: The Private Life of War* (New York: Anchor, 1993), 3, 15.

24. Robert Jay Lifton and Eric Markusen, *The Genocidal Mentality: Nazi Holocaust and Nuclear Threat* (New York: Basic, 1990).

25. Ibid., 13.

26. Ibid.

27. Robert Jay Lifton, *The Nazi Doctors: Medical Killing and the Psychology of Genocide* (New York: Basic, 1986).

28. Ibid., 319.

29. C. G. Jung, *After the Catastrophe*, in *The Collected Works of C. G. Jung*, vol. 10, trans. R. F. C. Hull, ed. Herbert Edward Read, Michael Fordham, Gerhard Adler (New York: Routledge, 1953), 198; Murray Stein, ed., *Jung on Evil* (Princeton: Princeton University, 1995).

30. Jung, *After the Catastrophe,* 199.

31. Lifton, *Nazi Doctors,* 334.

32. Edward S. Herman, *The Triumph of the Market: Essays on Economics, Politics, and the Media* (Boston: South End, 1999), 98.

33. Tony Judt, "The 'Problem of Evil' in Post-War Europe," *New York Review of Books,* February 14, 2008, http://www.nybooks.com/articles/21031.

34. Peter Nabokov, *Native American Testimony: A Chronicle of Indian and White Relations from Prophecy to the Present, 1492–1992* (New York: Penguin, 1992), 181–82, 184.

35. Quoted in Charles Patterson, *Eternal Treblinka: Our Treatment of Animals and the Holocaust* (New York: Lantern, 2002); Ray Ryan, personal communication, December 2, 2008.

36. Ros Clubb et al., "Compromised Survivorship in Zoo Elephants," *Science* 322 (2008): 1649.

37. See, for example, the Convention Against Torture and Other Cruel, Inhumane or Degrading Treatment or Punishment, United Nations, December 10, 1984: "Any act by which severe pain or suffering, whether physical or mental, is intentionally inflicted on a person for such purposes as obtaining from him, or a third person, information or a confession, punishing him for an act he or a third person has committed or is suspected of having committed, or intimidating or coercing him or a third person, or for any reason based on discrimination of any kind, when such pain or suffering is inflicted by or at the instigation of or with the consent or acquiescence of a public official or other person acting in an official capacity. It does not include pain or suffering arising only from, inherent in, or incidental to, lawful sanctions"; http://www.hrweb.org/legal/cat.html.

38. Lifton, *Nazi Doctors,* 167.

39. Jung, *After the Catastrophe,* 201.

40. C. G. Jung, *Civilization in Transition,* in *Collected Works,* 10: 618; C. G. Jung, *Two Essays on Analytical Psychology,* ibid., 7 (1953): 153.

41. Quoted in Patterson, *Eternal Treblinka,* 140.

42. Lisa Peattie, "Normalizing the Unthinkable," *Bulletin of Atomic Scientists,* March 1984.

43. Herman, *Triumph of the Market.*

44. Ibid.; Hannah Arendt, *Eichmann in Jerusalem: A Report on the Banality of Evil* (New York: Penguin Classics, 1992), 97.

45. Lifton, *Nazi Doctors,* 12.

46. Randy Malamud, personal communication, January 5, 2009.

47. Interim Hearing.

48. Kathy Lynn Gray, "New Executive Director Takes Helm at Zoo," *Columbus*

Dispatch, July 6, 2008, http://dispatch.com/live/content/local_news/stories/2008/07/06/newzooguy.html?sid=101.

49. "Activists Hope Dallas Zoo Elephant Gets New Home For B'Day," January 2, 2009, http://www.nbcdfw.com/news/local/Activist-Group-Hopes-Jenny-The-Elephant-Gets-New-Home-For-Her-Birthday-.html.

50. Mike di Paola, "Zoo-Bred Elephants Won't Help Save 30,000 in Wild: Commentary," June 26, 2008, http://www.bloomberg.com/apps/news?pid=20601088&sid=aGyZz7y.obTU&refer=muse.

51. Thomas B. Hildebrandt et al., "Ultrasonography of the Urogenital Tract in Elephants (*Loxodonta africana* and *Elephas maximus*): An Important Tool for Assessing Male Reproductive Function," *Zoo Biology* 19 (2000): 333–45.

52. Jesse Donahue and Erik Trump, *The Politics of Zoos: Exotic Animals and Their Protectors* (DeKalb: Northern Illinois University Press, 2006), 109; Michael Hutchins and Mike Keele, "Elephant Importation from Range Countries: Ethical and Practical Considerations for Accredited Zoos," *Zoo Biology* 25 (2006): 219–33.

53. Matthew Scully, *Dominion: The Power of Man, the Suffering of Animals, and the Call to Mercy* (New York: Macmillan, 2003).

54. Interim Hearing.

55. "Ringling Bros. and Barnum & Bailey: 139 Years of Tradition, Excitement, and Wonder," http://www.feldentertainment.com/pr/presskit/RinglingBros.Profile.pdf.

56. Michael Schmidt, *Jumbo Ghosts: The Dangerous Life of Elephants in the Zoo* (Philadelphia: Xlibris, 2002), 19.

57. Lynn D. Dierking et al., *Visitor Learning in Zoos and Aquariums: Executive Summary and a Literature Review* (Silver Spring, Md.: AZA, 2001–2); quoted in Lisa Kane, "Contemporary Zoo Elephant Management: Captive to a 19th-Century Vision," in *An Elephant in the Room: The Science and Well-Being of Elephants in Captivity*, ed. Debra L. Forthman et al. (North Grafton, Mass.: Tufts Center for Animals and Public Policy, 2009), 87–98.

58. John H. Falk et al., "Why Zoos and Aquariums Matter: Assessing the Impact of a Visit to a Zoo or Aquarium," Association of Zoos and Aquariums, 2007, http://www.aza.org/ConEd/Documents/Why_Zoos_Matter.pdf.

59. Lori Marino et al., "Do Zoos and Aquariums Promote Attitude Change in Visitors? A Critical Evaluation of the American Zoo and Aquarium Study," *Society and Animals*, in press.

60. Anne Baker, quoted in Kane, "Contemporary Zoo Elephant Management," 94.

61. Ibid., 95.

62. Ray Ryan, personal communication, December 2, 2008.

63. See Roxanna Carrillo et al., "Not a Minute More: Ending Violence Against

Women," United Nations Development for Women, 2003, http://www.unifem.org/resources/item_detail.php?ProductID=7.

64. Virgil Butler, "More of My Personal Story, Part Two," The Cyberactivist, August 6, 2005, http://cyberactivist.blogspot.com/2005/08/more-of-my-personal-story-part-two.html.

65. American Humane Association, "Violence to Humans and Animals: An Important Link," June 22, 2001, www.americanhumane.org/children/factsheets/viol_link.htm; Charolotte A. Lacroix, "Another Weapon for Combating Family Violence: Prevention of Animal Abuse," in *Child Abuse, Domestic Violence, and Animal Abuse: Linking the Circles of Compassion for Prevention and Intervention,* ed. Frank R. Ascione and Phil Arkow (West Lafayette, Ind.: Purdue University Press, 1999), 65.

66. Interim Hearing, 53.

67. Ray Ryan, personal commmunication, December 2, 2008. Ryan's description is reminiscent not only of James Gilligan's prisoners who are "dead inside" but also of what Henry Krystal called "robotization" observed in camp prisoners. See Henry Krystal, "Trauma and Affects," *Psychoanalytic Study of the Child* 33 (1978): 81–116.

68. Herman, *Triumph of the Market,* 99.

69. Ibid., 75.

70. Doug Meyers, internal memo to San Diego Zoological Society employees, March 26, 1989.

71. Lifton, *Nazi Doctors,* 229.

72. Herman, *Triumph of the Market,* 99.

73. A. J. Wilson, Letter to Mr. Douglas Myers, Executive Director, April 3, 1989, APHIS.

74. San Diego City Attorney John Witt, news release, June 30, 1988.

75. "Patricia Sheridan's Breakfast with . . . Willie Theison," *Post-Gazette NOW,* August 18, 2008, http://www.post-gazette.com/pg/08231/905081–42.stm; Michael Hutchins, "Variation in Nature: Its Implications for Zoo Elephant Management," *Zoo Biology* 25 (2006): 161–71.

76. Rich Lord, "Herd Mentality," *Pittsburgh Magazine,* February 2004, http://www.wqed.org/mag/features/0204_elephant_whisperer.shtml; "Team 4: Elephant Was Exposed to Violence," Pittsburgh Channel, November 21, 2002, http://www.thepittsburghchannel.com/team4/1800419/detail.html.

77. Quoted in Lifton, *Nazi Doctors,* 165.

78. Ibid., 425.

79. Herman, *Triumph of the Market,* 99.

80. Francis Jeanson, "Cette Algérie conquise et pacifiée," *Esprit,* April 1950, 624, quoted in Frantz Fanon, *Black Skin, White Masks,* trans. Richard Philcox (New York: Grove, 2008), 72.

81. Jung, "Epilogue," 198.

82. Edward Tick, *War and the Soul: Healing Our Nation's Veterans from Post-Traumatic Stress Disorder* (Wheaton, Ill.: Quest, 2005), 168.

83. Elke Riesterer, personal communication, July 3, 2007.

12. Beyond Numbers

1. "Elephant Left to Starve Dies," *Hindu*, March 18, 2009, http://www.thehindu.com/2009/03/18/stories/2009031855750800.htm.

2. Bruce Wilshire, *Get 'Em All! Kill 'Em!* (Lanham, Md.: Lexington, 2006), xviii.

3. Quoted in Shankar Vedantam, "How Brain's 'Mirrors' Aid Our Social Understanding," *Washington Post*, September 25, 2006, http://vedantam.com/mirror-09-2006.html; Arthur Glenburg, personal communication, January 16, 2009.

4. Caroline Preston, "Americans Are Passionate About Social Causes, But Few Take Action Based on Their Beliefs, Study Finds," *Chronicle of Philosophy*, February 6, 2009, http://philanthropy.com/news/updates/7054/americans-are-passionate-about-social-causes-but-few-take-action-based-on-their-beliefs-study-finds.

5. See Chris Mooney, *The Republican War on Science* (New York: Basic, 2005); Todd Wilkinson, *Science Under Siege: The Politician's War on Nature and Truth* (Boulder, Colo.: Johnson, 1998).

6. Michael Wines, "South Africa Considers Culling Elephants as Last Resort," *New York Times*, March 1, 2007, http://www.nytimes.com/2007/03/01/world/africa/01safrica.html?_r=1&oref=slogin&pagewanted=print.

7. Policy Announcement by Marthinus van Schalkwyk, South African Minister of Environmental Affairs and Tourism, on the Occasion of the Publication of the Final Norms and Standards for Elephant Management, Pretoria, February 25, 2008, http://www.info.gov.za/speeches/2008/08022516451001.htm.

8. Robert J. Scholes et al., Summary for Policymakers: Elephant Management in South Africa, October 15, 2007, www.elephantassessment.co.za/files/SPMv03RJS.pdf, 5.

9. *Assessment of South African Elephant Management 2007*, http://www.elephantassessment.co.za/files/Assessment-of-South-African-Elephant-Management-2007.pdf, 133.

10. Robert J. Scholes, Assessment of South African Elephant Management, Powerpoint presentation to media representatives, February 25, 2008, slide 10.

11. T. M. Palmer et al., "Breakdown of an Ant-Plant Mutualism Follows Loss of Large Herbivores from an African Savannah," *Science* 319 (2008): 192–95.

12. *Assessment of South African Elephant Management 2007*, 132.

13. Ibid., 152, 167.

14. Scholes et al., Summary for Policymakers, 5.

15. Policy Announcement.

16. *Assessment of South African Elephant Management 2007*, ii.

17. Beatrice Obwocha and Winnie Chumo, "Birth Control for Elephants to Start (Kenya)," *East African Standard,* February 2, 2007.

18. Rudi J. van Aarde and Tim P. Jackson, "Megaparks for Metapopulations: Addressing the Causes of Locally High Elephant Numbers in Southern Africa," *Biological Conservation* 134 (2007): 289–97.

19. See Amboseli Elephant Research Project (AERP), Statement on Elephant Culling, www.elephanttrust.org; Animal Rights Africa, www.animalrightsafrica.org.

20. AERP, Statement on Elephant Culling.

21. *Assessment of South African Elephant Management 2007,* 155.

22. Thidinalei Tshiguvho, "Sacred Traditions and Biodiversity Conservation in the Forest Montane Region of Venda, South Africa," Ph.D. diss., Clark University, 2008.

23. *Assessment of South African Elephant Management 2007,* 148.

24. Tshiguvho, "Sacred Traditions and Biodiversity Conservation."

25. "Animal Protectionists, Conservationists, and Civil Society Up in Arms at van Schalkwyk's Apparent Contempt for Prescribed Public Consultation Processes," Animal Rights Africa, press release, February 2, 2009, http://www.animalrightsafrica.org/PR_02Feb09.php.

26. Mike Cadman, personal communication, March 3, 2009.

27. Stewardship of federal lands in the United States is heavily influenced by military heritage; proponents of the National Forest Commission Report that led to legislature governing forest reserve use (Organic Act of 1897) maintained that forest officers' duties were "military in character and should be regulated . . . on military principles"; Gifford Pinchot, *Breaking New Ground,* commemorative ed. (Washington, D.C.: Island, 1998), 121.

28. *Assessment of South African Elephant Management 2007,* 414.

29. Scholes et al., Summary for Policymakers, 14–15.

30. Ibid., 8.

31. Ibid., 5.

32. *Assessment of South African Elephant Management 2007,* 296.

33. G. Wittemyer, I. Douglas-Hamilton, and W. M. Getz, "The Socio-ecology of Elephants: Analysis of the Processes Creating Multitiered Social Structures," *Animal Behaviour* 69 (2005): 1357–71.

34. *Assessment of South African Elephant Management 2007,* 291.

35. See G. A. Bradshaw, Allan N. Schore, Janine Brown, Joyce Poole, and Cynthia Moss, "Elephant Breakdown," *Nature* 433 (2005): 807; G. A. Bradshaw and Allan N. Schore, "How Elephants Are Opening Doors: Developmental Neuroethology, Attachment, and Social Context," *Ethology* 113 (2007): 426–36.

36. Lance H. Gunderson, C. S. Holling, and Stephen S. Light, eds., *Barriers and*

Bridges to the Renewal of Ecosystems and Institutions (New York: Columbia University Press, 1995).

37. AERP, Statement on Elephant Culling.

38. Scholes et al., Summary for Policymakers, 15.

39. *Assessment of South African Elephant Management 2007*, table 8.3, 286, citing C. C. Grant, compiler, "Elephant Effects on Biodiversity: An Assessment of Current Knowledge and Understanding as a Basis for Elephant Management in SANParks," SANParks Scientific Report, March 2005.

40. Ibid., 287.

41. Jessica Aldred, "Controversial Elephant Ivory Auction Begins in Southern Africa: China and Japan Bid for More Than 100 Tonnes of Stockpiled Elephant Ivory from South Africa, Namibia, Botswana and Zimbabwe," *Guardian*, October 28, 2008, http://www.guardian.co.uk/environment/2008/oct/28/wildlife-conservation.

42. "Botswana: On the Trunk Road," Telegraph.co.uk, March 30, 2001, http://www.telegraph.co.uk/travel/717661/On-the-trunk-road—Christopher-Munnion-tells-the-story-of-Randall-Moore-the-pioneer-of-elephant-back-safaris.html.

43. *Assessment of South African Elephant Management 2007*, 347ff; Mike Cadman, "Hunters Take Out Kruger's Animals: Activists Question the Government's Commitment to Protecting National Heritage," *Sunday Independent*, May 10, 2009.

44. Steve Best, "The Killing Fields of South Africa: Eco-Wars, Species Apartheid, and Total Liberation," 2007, http://www.drstevebest.org/Essays/TheKillingFields.htm.

45. Samuel K. Wasser et al., "Combating the Illegal Trade in African Elephant Ivory with DNA Forensics," *Conservation Biology* 22 (2008): 1065–71.

46. Subhash Chandra, "One Elephant Dies Every Third Day (India)," *Deccan Herald*, http://www.deccanherald.com/Content/Feb22009/scroll20090202116066.asp?section=updatenews.

47. A. Choudhury at al., "Elephas maximus," in 2008 IUCN Red List of Threatened Species, www.iucnredlist.org. The Asian elephant is "listed as Endangered (EN) because of a population size reduction inferred to be at least 50% over the last three generations, based on a reduction in its area of occupancy and the quality of its habitat. Although there are few accurate data on historical population size, from what is known about trends in habitat loss/degradation and other threats including poaching, an overall population decline of at least 50% over the last three generations (estimated to be 60–75 years, based on a generation time estimated to be 20–25 years) seems realistic."

48. J. Blanc, "Loxodonta africana," ibid. "The African Elephant is listed as Near Threatened on the basis of an inferred decline of 25% between 1979 and 2007, as this falls short of the 30% threshold required for a Vulnerable listing under criterion

A2a. The current assessment therefore represents a downlisting of the species with respect to its previous listing as Vulnerable (VU A2a) in the 2004 IUCN Red List."

49. Keith Lindsay, personal communication, January 16, 2009.

50. AERP, Statement on Elephant Culling.

51. Umli Miuli, "Elephants Face Death by Electrocution," IT Examiner.com, November 4, 2008, http://www.itexaminer.com/elephants-face-death-by-electrocution.aspx.

52. Matthew Scully, *Dominion: The Power of Man, the Suffering of Animals, and the Call to Mercy* (New York: Macmillan, 2003).

53. Jane Carruthers, "Wilding the Farm or Farming the Wild? The Evolution of Scientific Game Ranching in South Africa from the 1960s to Present," *Transactions of the Royal Society of South Africa* 63 (2008): 160–81.

54. *Assessment of South African Elephant Management 2007*, 26.

55. Quoted in Carruthers, "Wilding the Farm," 161; "Animal Warehouse: Park Builds Horde of Exotic Treasures," *Wall Street Journal*, September 19, 2007, http://online.wsj.com/article/SB119014975897631500.html.

56. Richard Leakey, "Is Culling Imminent for South African Elephants?" *Wildlife Direct*, February 28, 2008, http://richardleakey.wildlifedirect.org/2008/02/28/is-culling-imminent-for-south-african-elephants/.

57. International Fund for Animal Welfare, "Kruger National Park: Elephant Cull or Killing Fields?" http://www.ifaw.org/ifaw_southern_africa/join_campaigns/national_regional_efforts/the_debate_on_elephant_culling_in_south_africa/index.php.

58. "Conservation Group Sees Need for Elephant Cull (South Africa)," *New Scientist*, February 26, 2007, http://environment.newscientist.com/article/dn11258-conservation-group-sees-need-for-elephant-cull.html.

59. Quoted in Virginia Morell, "Do Zoos Shorten Elephant Life Spans?" *Science NOW Daily News*, December 11, 2008; Oakland Zoo, "In the Field: Africa," http://www.oaklandzoo.org/conservation-programs/in-the-field-africa.

60. Quoted in Charles Patterson, *Eternal Treblinka: Our Treatment of Animals and the Holocaust* (New York: Lantern Books, 2002), 206.

61. Tom Regan, *The Case for Animal Rights* (Berkeley: University of California Press, 2004); Patterson, *Eternal Treblinka*, 207.

62. John Gluck, personal communication, January 13, 2009.

63. John E. Mack, "The Politics of Species Arrogance," in *Ecopsychology: Restoring the Earth, Healing the Mind*, ed. Theodore Roszak, Mary E. Gomes, and Allen D. Kane (San Francisco: Sierra Club Books, 1995), 279–87.

64. David Joravsky, "The Scientist as Conformist," *New York Review of Books*, October 12, 1978, quoting Albert Einstein and Max Born, *Briefwechsel* (Munich: Nymphenburger Verlagshandl, 1969), 203.

65. Robert Jay Lifton, *The Nazi Doctors: Medical Killing and the Psychology of Genocide* (New York: Basic, 1986), 418–29.

66. Joravsky, "Scientist as Conformist."

67. Scholes et al., Summary for Policymakers, 5.

68. Roy Harvey Pearce, *Savagism and Civilization: A Study of the Indian and the American Mind* (Berkeley: University of California Press, 1988). 64.

69. Brian W. Dippie, *The Vanishing American: White Attitudes and U.S. Indian Policy* (Middletown, Conn.: Wesleyan University Press, 1982), xv.

70. Susan Scheckel, *The Insistence of the Indian: Race and Nationalism in Nineteenth-Century American Culture* (Princeton: Princeton University Press, 1998), 76.

71. "Belonged in the American past": Pearce, *Savagism and Civilisation,* 160.

Epilogue

1. Phoebe G. Linden, personal communication, October 17, 2008.

2. Kai J. Erikson, *A New Species of Trouble: The Human Experience of Modern Disasters* (New York: Norton, 1994), 11.

3. Kelly Oliver, *The Colonization of Psychic Space: A Psychoanalytic Social Theory of Oppression* (Minneapolis: University of Minnesota Press, 2004), 199.

4. C. G. Jung, *Civilization in Transition,* in *The Collected Works of C. G. Jung,* vol. 10, trans. R. F. C. Hull, ed. Herbert Edward Read, Michael Fordham, Gerhard Adler (New York: Routledge, 1953), 618.

5. T. E. Lawrence, *Seven Pillars of Wisdom* (Hertfordshire, U.K.: Wordsworth Editions, 1997), 7.

6. *Assessment of South African Elephant Management 2007,* http://www.elephantassessment.co.za/files/Assessment-of-South-African-Elephant-Management-2007.pdf, 309.

7. Patricia Hill-Collins, "The Meaning of Motherhood in Black Culture and Black Mother-Daughter Relationships," in *Double Stitch: Black Women Write about Mothers and Daughters,* ed. Patricia Bell-Scott (Boston: Beacon 1991), 42–62.

8. Beverly Guy-Sheftall, "Piecing Blocks: Identities," in Bell-Scott, *Double Stitch,* 61–62.

Index

Index